Science
Progress 1

Andrea Coates
Michelle Austin
Richard Grimmer

DYNAMIC
LEARNING

HODDER
EDUCATION
AN HACHETTE UK COMPANY

Although every effort has been made to ensure that website addresses are correct at time of going to press, Hodder Education cannot be held responsible for the content of any website mentioned. It is sometimes possible to find a relocated web page by typing in the address of the home page for a website in the URL window of your browser.

Orders: please contact Hachette UK Distribution, Hely Hutchinson Centre, Milton Road, Didcot, Oxfordshire, OX11 7HH. Telephone: +44 (0)1235 827827. Email education@hachette.co.uk Lines are open from 9 a.m. to 5 p.m., Monday to Friday. You can also order through our website: www.hoddereducation.co.uk

© Andrea Coates, Michelle Austin, Richard Grimmer and Mark Edwards, Sue Hocking, Bevely Rickwood 2014

First published in 2014 by
Hodder Education
An Hachette UK Company,
Carmelite House,
50 Victoria Embankment London, EC4Y 0DZ

Impression number   15

Year   2023

Cover photo © Jeffrey Collingwood – Fotolia

Typeset in 11.5/13 ITC Officina Sans by Aptara, Inc.

Printed and bound by CPI Group (UK) Ltd, Croydon, CR0 4YY

A catalogue record for this title is available from the British Library.

ISBN 978 1 4718 0142 6

Contents

→ CHEMISTRY

→ **PHYSICS**

Get the most from this book

Welcome to Science Progress Student's Book 1!

This book covers the first half of your KS3 Science course, divided into Biology, Chemistry and Physics sections, each covering 6 Topics.

As you work through the year, you will do a combination of the Biology, Chemistry and Physics Topics so you can use the coloured tabs on the right hand side of the pages to easily find the right Topic.

We hope you will enjoy this book as much as our authors have enjoyed creating the content, questions and activities for you!

→ Learn more, show your understanding, and build scientific enquiry skills

This book has been carefully designed to help you build your KS3 Science knowledge, understanding and skills.

Start with an interesting **Context** to see where this topic fits into the real world and the rest of your Science course. See if you can answer the question.

Answer the questions in the photo captions to practice thinking like a scientist. This is called "scientific enquiry".

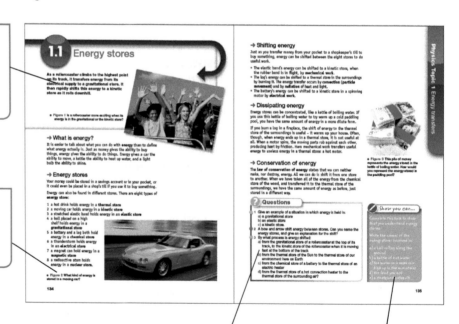

Work through the **Questions** to test your knowledge and understanding. The questions are colour-coded from simple to more advanced so that you can monitor your own progression. Challenge yourself and answer the blue questions.

Complete the **Show you can** task to show that you are confident in your understanding of this topic and that you have learnt the right scientific skills.

→ Build Working Scientifically skills

In this book, Working Scientifically skills are covered in the short questions throughout and explored in more detail in the activity at the end of each Topic.

Work through these activities to build and practice Working Scientifically skills such as Planning and Designing Investigations, Calculating Scientifically and Presenting and Interpreting Data.

Each activity includes a scenario for you to work through with questions to guide you in completing the task. You can complete these on your own or in groups.

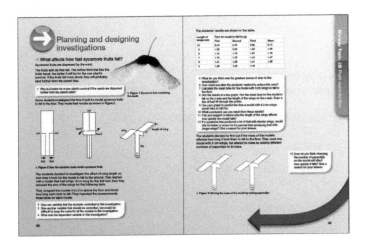

→ Free online extras to help you build your knowledge and understanding

At the back of this book you will find QR codes. You can use a free mobile app to open the online extras. You can also type in the short weblink to find the same content.

- For each set of Questions, you will find Hints to help you answer the questions.
- For all Key Terms, find more detailed explanations and examples in the Extended Glossary.
- Once you are confident that you have learnt all the key words in the Topic, complete the Key Words Tests online.

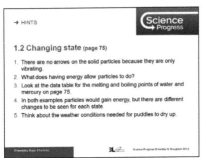

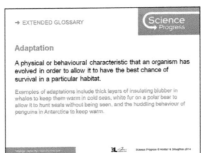

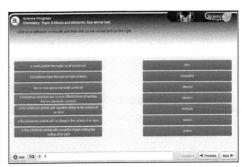

1.1 Microscopes

Cells are the building blocks of all living things, but they are so small that we can only see and study them using a microscope. When microscopes were invented they also allowed us to see tiny organisms that previously had been completely invisible to us.

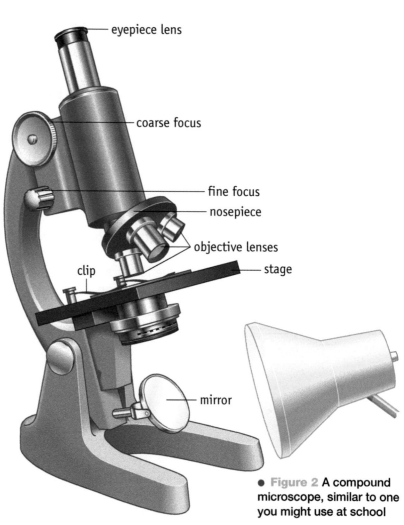

● **Figure 1** Organisms in pond water that can only be seen under a microscope. How many different types can you see?

In the late 17th century, the English scientist Robert Hooke used a compound microscope (with two lenses) to look at very small objects. The two lenses meant that his microscope could magnify objects much more than a single lens. Hooke was the first person to use the word 'cell'.

Figure 2 shows a compound microscope. In this microscope there are two lenses. These kinds of microscope are sometimes called light microscopes because they use light to let you see things.

→ How to use a compound microscope to view a prepared slide

Wear eye protection.

1 Place the slide on the stage of the microscope so that the **specimen** is directly above the hole in the stage. Clip it in place.
2 Adjust the position of the mirror so that light reflected from the lamp passes up through the specimen.
3 Rotate the nosepiece so that the smallest (lowest power) objective lens is directly over the specimen.

eyepiece lens

coarse focus

fine focus

nosepiece

objective lenses

clip

stage

mirror

● **Figure 2** A compound microscope, similar to one you might use at school

4 Turn the coarse focus knob to bring the objective lens close to the stage.

5 Use one eye to look down the eyepiece. Slowly turn the coarse focus knob to focus the image. Take care not to move the objective lens too far down or you will crack the slide.

6 Use the fine focus knob to get a clear, sharp image.

7 If you want to use a higher magnification, make sure that the object is in the centre of your field of view. Rotate the next size objective lens into place. Readjust the focus using the fine focus knob.

→ How to prepare your own slide

You can prepare your own slides – follow the procedure in Figure 3. Remember to wear eye protection. Make sure the sample of **tissue** is thin enough to allow light to pass through it. You can mount your specimen in water, but using a stain makes it easier to the see the structures inside the cells.

→ Magnification

Magnification is the number of times greater the size of the image is compared to the size of the object. If something is magnified ×40, it means that the image you see with a microscope is 40 times bigger than the actual object. Both the eyepiece lens and the objective lens magnify the object.

The formula for magnification is:

total magnification = eyepiece lens magnification × objective lens magnification

If the eyepiece lens has a magnification of ×10 and the objective lens has a magnification of ×4, then the total magnification = $10 \times 4 = 40$. We write the magnification on drawings made from microscope images. This magnification would be written as ×40.

Put a drop of stain onto a clean microscope slide.

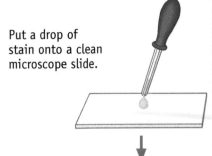

Place a thin piece of your specimen onto the stain, making sure it stays flat. You might want to add another drop of stain on top of the specimen.

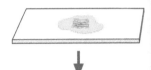

Use a mounted needle to carefully lower a coverslip onto the specimen. If you do this too fast you may trap air bubbles.

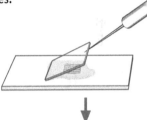

Use a piece of tissue paper to mop up any excess stain that is outside the coverslip. You can now look at your slide using a microscope.

● Figure 3 The main steps in preparing a microscope slide

? Questions

1 Why do we need to use a microscope to see cells?

2 A slide is viewed under a microscope using an eyepiece lens with a magnification of ×5 and an objective lens with a magnification of ×20. Calculate the total magnification.

3 Why do scientists often use stains when preparing microscope slides?

4 Suggest why a compound microscope magnifies things more than a magnifying glass.

5 **Electron microscopes** are much more powerful than light microscopes. We can use them to see viruses, like the cold virus, that are much smaller than cells. Find out about the main differences between a light microscope and an electron microscope. Write a half-page report on your findings. You might want to include a picture.

✎ Show you can...

Complete this task to show that you understand how to use a microscope.

Describe step by step how to view a prepared slide under high power.

1.2 Cells

Some organisms are adapted to survive in extreme conditions. Watermelon snow is a micro-organism that lives in frozen water. Inside its cell is a red pigment. If you walk on snow where these cells live, the snow gets stained red when the cells get damaged. The blood-red stains on snow puzzled explorers for hundreds of years.

● Figure 1 A watermelon snow cell. How do you think scientists worked out what caused the red stains on snow?

Cells are the building blocks of living organisms. Some living organisms consist of only one cell. They are described as **unicellular organisms.** All bacteria are unicellular. **Multicellular organisms** are made up of many cells. Plants and animals are made up of billions of cells.

→ Cell structure

Cells are the smallest parts of living organisms. There are smaller structures inside a cell, each with its own function.

Animal cells

Look at the structures in the human cheek cell. Most cells have these structures.

a)

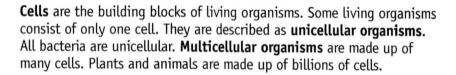
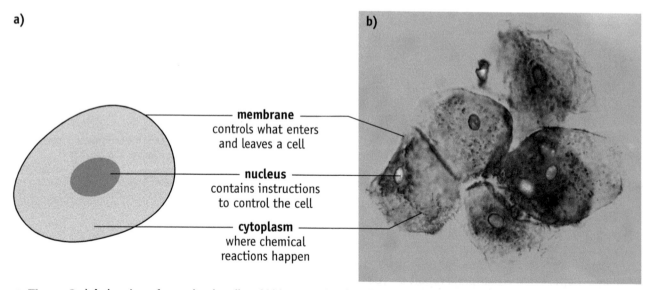

membrane
controls what enters
and leaves a cell

nucleus
contains instructions
to control the cell

cytoplasm
where chemical
reactions happen

b)

● Figure 2 a) A drawing of one cheek cell and b) human cheek cells magnified about 1000 times. How big do you think a cheek cell is?

Plant cells

Plant cells are usually bigger than animal cells. They have a **nucleus**, a **cell membrane** and **cytoplasm** like animal cells, but they also have some other structures, such as **chloroplasts**. Inside the chloroplasts is a green substance called **chlorophyll**. This absorbs light energy for photosynthesis.

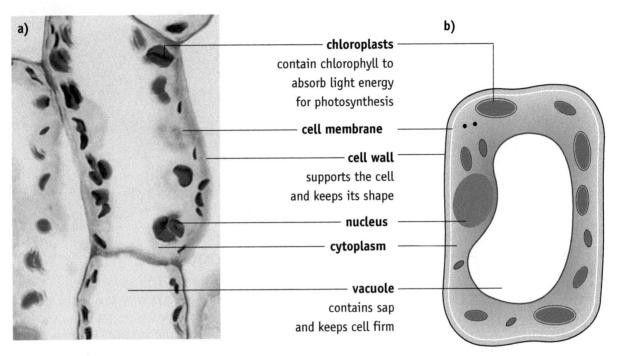

a)

b)

chloroplasts
contain chlorophyll to
absorb light energy
for photosynthesis

cell membrane

cell wall
supports the cell
and keeps its shape

nucleus

cytoplasm

vacuole
contains sap
and keeps cell firm

● Figure 3 a) Plant cells magnified about 1000 times and b) a drawing of one plant cell. How many times bigger than a human cheek cell are these plant cells?

Both plant and animal cells also have small structures in the cytoplasm called **mitochondria**. **Respiration** happens in the mitochondria. Respiration releases energy from food. Cells that need a lot of energy, such as muscle cells, have many mitochondria.

? Questions

1 Bacteria are unicellular organisms. What does 'unicellular' mean?
2 Describe what a multicellular organism is.
3 Name four structures found in both plant and animal cells.
4 Which structures are found in plant cells but not in animal cells?
5 Muscle cells contain many mitochondria. Explain why.

Show you can...

Complete this task to show that you know the structures in plant and animal cells.

Copy and complete the table to name all the cell structures. Tick whether they are found in plant cells, animal cells, or both. Explain what their function is in the cell. The first line has been done for you.

Structure	Plant	Animal	Function
Cytoplasm	✓	✓	Where chemical reactions happen

1.3 Specialised cells

Not all cells look the same. They have different shapes because they are specialised for a specific function. Histologists are scientists who study cells under the microscope. They identify cells and diagnose diseases.

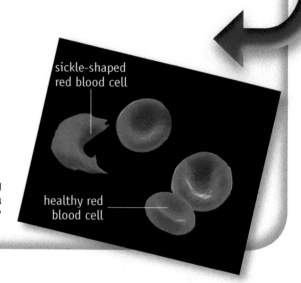

sickle-shaped
red blood cell

healthy red
blood cell

● **Figure 1** These red blood cells are from someone suffering from a disease called sickle cell anaemia. What would a histologist be looking for to diagnose sickle cell anaemia?

There are many different types of cell in the human body, each one **specialised** for its own particular function.

→ Nerve cells

Nerve cells are specialised for carrying electrical impulses over a long distance. They can do this as they are very long.

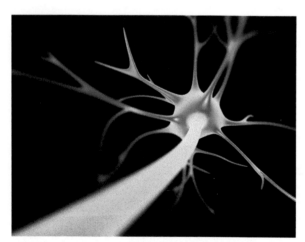

● **Figure 2** A nerve cell. How long do you think human nerve cells can be?

→ Red blood cells

Red blood cells are specialised for carrying oxygen around the body. They are very small and round, with a dimple in the middle, so they can squeeze through the tiniest blood vessels. They do not have a nucleus, which leaves more space to carry a lot of oxygen.

→ Skin cells

Skin cells prevent microbes from entering the body. They form a smooth layer as they are very flat and closely packed together.

● **Figure 3** Skin cells. How many skin cells do you think we lose every day?

→ Leaf cells

Plant cells are also specialised for different functions. Cells in a plant leaf are specialised to make food by **photosynthesis**. They contain lots of chloroplasts. Chloroplasts contain the green pigment called chlorophyll that absorbs the light energy needed for photosynthesis.

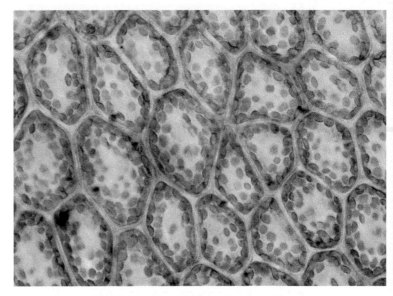

● **Figure 4** Leaf cells. What makes leaves look green?

→ Root hair cells

Root hair cells are specialised for absorbing water from the soil. The long root hair has a large surface area and absorbs a lot of water.

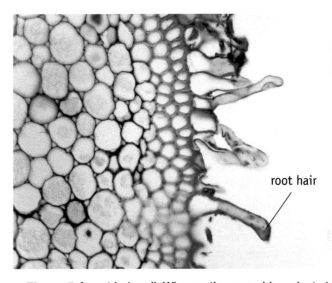

root hair

● **Figure 5** A root hair cell. Why are there no chloroplasts in root hair cells?

? Questions

1 Describe what histologists do.
2 Describe how skin cells are specialised for their function in the body.
3 Explain how plant cells are specialised to make food.
4 Look at the photograph of the red blood cells from someone suffering from sickle cell anaemia on page 10.
  a) Suggest how the shape of the red blood cells will affect how they function.
  b) List the symptoms you think a person with sickle cell anaemia will have.

✎ Show you can...

Complete this task to show that you understand how cells are adapted for their functions.

Draw diagrams of two different animal cells. Label any special features of the cells that help them to carry out their functions.

1.4 Simple and complex organisms

There are many simple organisms, such as bacteria, that are made up of only one cell. Other organisms, such as humans or trees, are much more complex and are made up of many different types of cell.

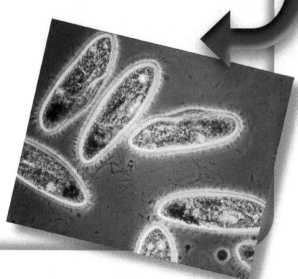

● Figure 1 *Paramecium cells*. Each single cell is about 0.2 mm long and covered in cilia (microscopic hair-like structures). What do you think is the function of these cilia?

→ *Amoeba* and *Euglena*

Amoeba and *Euglena* are unicellular organisms.

Amoeba lives in ponds. It moves along by pushing cytoplasm out into **pseudopodia** (false feet), which it uses to move towards its prey. Then it wraps the pseudopodia around the food. The food is trapped inside a food vacuole, where it is digested. Undigested material is released when the *Amoeba* moves on and opens the food vacuole to the water outside.

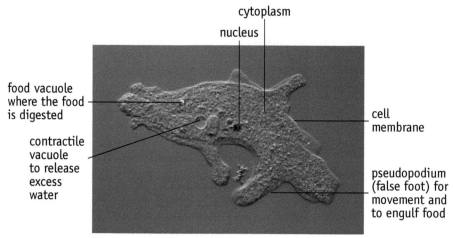

cytoplasm
nucleus
food vacuole where the food is digested
contractile vacuole to release excess water
cell membrane
pseudopodium (false foot) for movement and to engulf food

● Figure 2 *Amoeba*, about to engulf some food. It cannot see, so how do you think it detects its food?

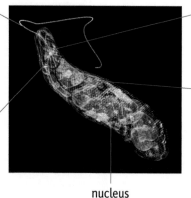

flagellum: used like a whip to move through the water
gullet: where food particles are taken in from the water
red eye spot: detects light
chloroplast: to make food by photosynthesis
nucleus

Euglena also lives in ponds and puddles. It swims using a whip-like **flagellum**. The cell contains chloroplasts, so *Euglena* can make food by photosynthesis. It detects where the light is using its red eye spot. It can also feed on particles in the water, which it takes into its gullet.

● Figure 3 *Euglena*. Why do you think it is important that *Euglena* can detect light?

Both *Amoeba* and *Euglena* take in oxygen from the water by a process called **diffusion**. Diffusion is the movement of a substance from an area where it is more highly concentrated to an area where it is less concentrated. There is a higher concentration of oxygen in the water than inside an organism, so oxygen diffuses into the organism. Carbon dioxide will diffuse in the opposite direction, out of the organism into the water. Diffusion is also important in multicellular organisms to move substances into and between cells.

→ Organisation in multicellular organisms

In large, multicellular organisms, most of the cells are too far away from the surface of the organism for chemicals to reach them just by diffusion. To get around this problem the cells in multicellular organisms are arranged in a precise way, as shown in the flow diagram, so all the cells are close to a blood capillary.

For example, muscle cells are grouped together to form muscle tissue that contracts and causes movement.

The stomach is an organ that is made up of several tissues including muscle tissue, nervous tissue and glandular tissue.

The stomach is one of the organs in the digestive system, which is one of the systems that makes up a human organism.

carbon dioxide diffuses out of the cell → lower concentration of carbon dioxide in the water

oxygen diffuses into the cell → higher concentration of oxygen in the water

● Figure 4 **Diffusion of gases into and out of a cell**

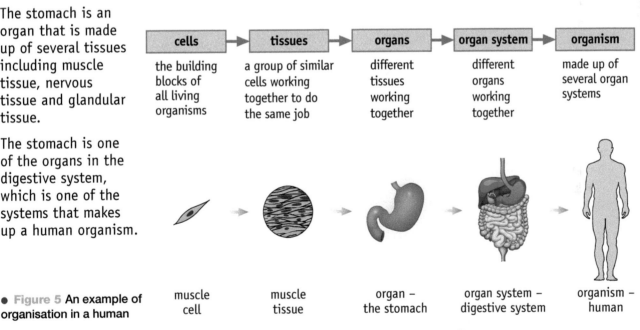

cells	→	tissues	→	organs	→	organ system	→	organism
the building blocks of all living organisms		a group of similar cells working together to do the same job		different tissues working together		different organs working together		made up of several organ systems

muscle cell → muscle tissue → organ – the stomach → organ system – digestive system → organism – human

● Figure 5 **An example of organisation in a human**

? Questions

1 Describe the function of the flagellum in *Euglena*.
2 When *Euglena* was first seen under a microscope some scientists thought it was a type of plant cell. Suggest why they thought this.
3 What is 'diffusion'?
4 Draw a flow diagram to show an example of how nerve cells are organised in the human body.

✎ Show you can...

Complete this task to show that you understand how *Amoeba* feeds.

Describe how *Amoeba* feeds. You could include drawings to illustrate your answer.

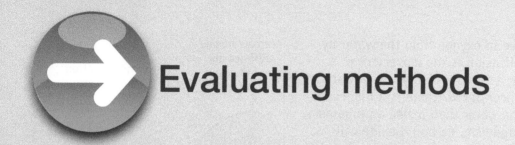

Evaluating methods

→ Looking at onion cells

A class of students was learning about cells. They prepared slides of onion cells to look at under the microscope. They used a drop of stain and put a put a thin piece of onion skin in it. Then they put a coverslip over the top.

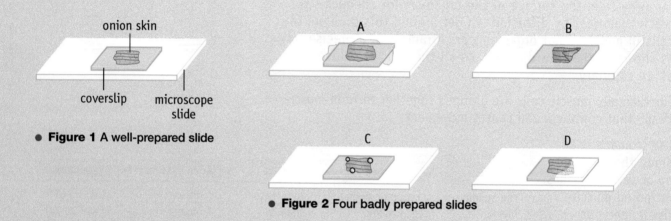

onion skin

coverslip · microscope slide

● **Figure 1** A well-prepared slide

A   B

C   D

● **Figure 2** Four badly prepared slides

1 Look at Figure 2. For each part, A, B, C and D, describe what is wrong with the slide.
2 For each slide explain how to avoid or correct the problem.

Measuring cells

Most cells are very small and only a few are big enough to be seen by the naked eye. *Amoeba* is a unicellular organism that lives in water. It is quite a large cell.

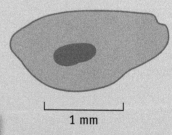

1 mm

● **Figure 3** *Amoeba*

3 **a)** Use the scale to estimate the width of an *Amoeba*.
  b) Do you think you could see an *Amoeba* by eye?

An egg cell is one of the largest cells found in the human body.

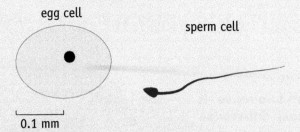

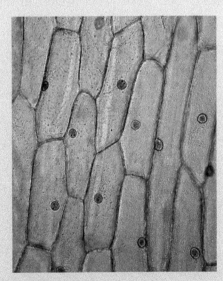

egg cell

sperm cell

0.1 mm

● **Figure 4** A human egg cell and a human sperm cell

4 Use the scale in Figure 4 to estimate the diameter of the egg cell and length of the sperm cell.

Using the scale given in Figure 4, the head of the sperm cell is about 0.03 mm long. For small cells it is not very convenient to measure the size in millimetres (mm). A smaller unit called the micrometre (µm) is used by scientists. 1 µm = 0.001 mm

To convert millimetres to micrometres you have to multiply by 1000. For example, something 0.2 mm long would measure $0.2 \times 1000 = 200$ µm.

5 How long is the head of a sperm cell in micrometres?

The photograph shows some plant cells.

6 a) One of the cells in Figure 5 measures about 40 µm by 15 µm. On graph paper draw a scale drawing showing the outline of one of these plant cells. You can draw it as a rectangle shape.
Remember to show the scale that you have used on the graph paper.
 b) Look up the size of other cells in a textbook or on the internet, and make scale drawings of these.
You could ask someone else to work out the actual size of the cells from your drawings.

● **Figure 5** Onion skin cells. How many of these cells do you think you could line up to cover 1 mm?

Male and female reproductive systems

In order for a particular type of organism to survive, it must be able to produce offspring (babies). Otherwise the species would die out. Most animal species have separate male and female organisms, but some, like snails, have both male and female reproductive systems in the same organism. They are said to be hermaphrodites.

● Figure 1 Two snails mating. How many fertilised eggs do you think these two snails will each carry after fertilisation?

→ The female reproductive system

When girls are born they have millions of **eggs** in each **ovary**. Once **puberty** begins, usually one egg is released from one ovary each month. This is called **ovulation**. If the egg is not fertilised the girl will have a **period**. This is when the womb lining breaks down and bleeds, because it is not needed to support a baby.

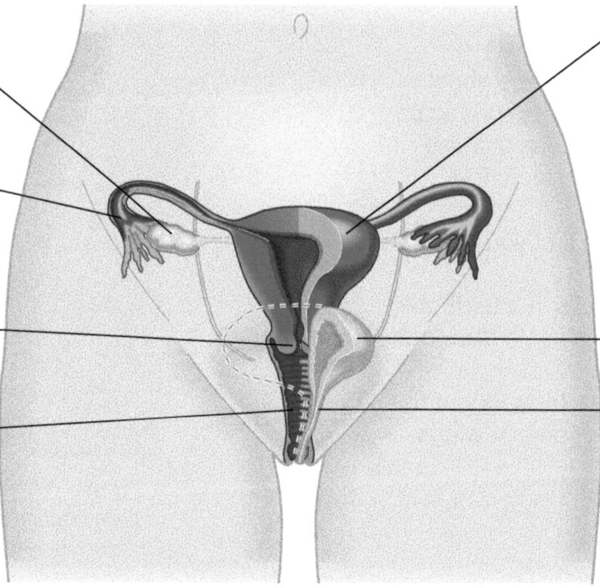

ovary – produces egg cells

oviduct – tube that carries an egg cell from the ovary towards the womb. Fertilisation occurs in the oviduct

cervix – the entrance to the womb

vagina – where sperm are deposited during sexual intercourse and where menstrual blood flows out of the body during a period

womb (uterus) – where a foetus (baby) develops. The womb lining breaks down each month if the woman is not pregnant

bladder – stores urine

urethra – tube that carries urine out of the body

● Figure 2 The female reproductive system

→ The male reproductive system

The **testes** (singular 'testis') are supported outside the body in the **scrotal sac**. This is because **sperm** production is more efficient at a temperature that is slightly lower than body temperature.

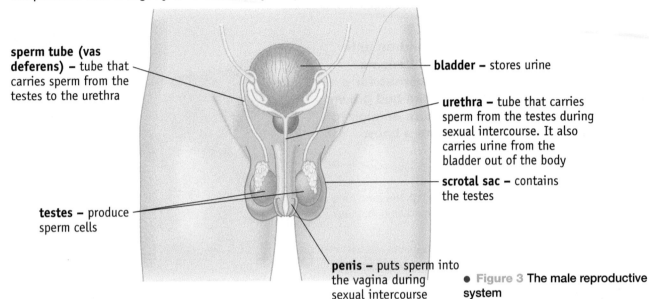

sperm tube (vas deferens) – tube that carries sperm from the testes to the urethra

bladder – stores urine

urethra – tube that carries sperm from the testes during sexual intercourse. It also carries urine from the bladder out of the body

scrotal sac – contains the testes

testes – produce sperm cells

penis – puts sperm into the vagina during sexual intercourse

● Figure 3 **The male reproductive system**

→ Infertility

Infertility is when a woman cannot get pregnant. There are various causes of infertility.

Some men have low sperm counts, which can be due to the testes being too warm. Wearing looser clothing can help to correct this problem. A man might produce damaged sperm that do not swim properly. His partner's eggs can be fertilised by sperm from a donor.

Some women do not ovulate regularly, or do not ovulate at all. This problem may be treated with **fertility drugs**. Eggs may be taken from a woman's ovaries, fertilised in a lab and inserted into her womb. This is called **IVF (in-vitro fertilisation)**. Alternatively, eggs from a donor can be fertilised by sperm from the woman's partner and then placed into her womb.

If a woman cannot carry a baby in her womb, her fertilised egg can be put into the womb of a **surrogate mother**. This raises ethical issues as the surrogate mother then has to give up a baby she has carried for nine months. In the future, women may be able to have a womb transplant.

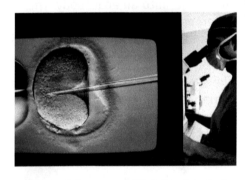

● Figure 4 **A sperm cell is being injected into an egg. IVF treatment is expensive and not always successful. Do you think this treatment should be offered to infertile couples by the NHS?**

Show you can...

Complete this task to show that you understand the male and female reproductive systems.

Describe some causes of infertility and suggest how the problems could be treated.

? Questions

1 Which organs in the female reproductive system produce eggs?
2 Which organs in the male reproductive system produce sperm?
3 Where would a baby grow inside a woman's body?
4 What are fertility drugs?
5 Surrogate mothers are women who agree to become pregnant with another couple's baby. Explain the ethical issues you think this raises.

2.2 Puberty and the menstrual cycle

Girls and boys cannot produce children until they have completed puberty and become sexually mature. It would be dangerous for a girl to become pregnant before she had grown sufficiently in size, and was mature enough to take care of herself and therefore a baby.

● **Figure 1** What do you think would be the problems if young children could have babies?

→ Puberty

Puberty is the time when a child's body starts to change into an adult body. This happens at different ages in different people, but generally starts between the ages of 9 and 14 in girls, and 11 and 15 in boys.

The changes happen because chemicals called **hormones** start to be produced in the body. In girls, the ovaries start to produce the hormone called **oestrogen**.

In boys the testes start to produce the hormone called **testosterone**. Several physical changes start to happen that make the body look different. These are called the **secondary sexual characteristics**.

Besides the physical changes, boys and girls also change emotionally and psychologically during puberty. These changes are also partly caused by the hormones produced.

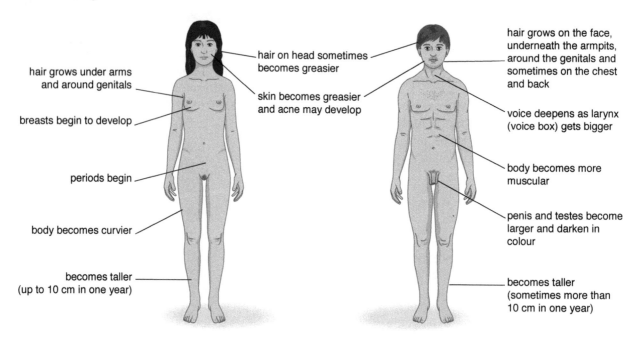

hair on head sometimes becomes greasier

hair grows under arms and around genitals

breasts begin to develop

skin becomes greasier and acne may develop

periods begin

body becomes curvier

becomes taller (up to 10 cm in one year)

hair grows on the face, underneath the armpits, around the genitals and sometimes on the chest and back

voice deepens as larynx (voice box) gets bigger

body becomes more muscular

penis and testes become larger and darken in colour

becomes taller (sometimes more than 10 cm in one year)

● **Figure 2 a)** Female secondary sexual characteristics and **b)** male secondary sexual characteristics

→ The menstrual cycle

From puberty until about the age of 50, girls and women have a monthly bleed from the vagina called a period or **menstruation**. It is part of the monthly cycle of changes called the **menstrual cycle**, which happens when the woman does not become pregnant.

The menstrual cycle lasts about 28 days. Day 1 of the cycle is the first day of menstruation. The lining of the **womb** (**uterus**) breaks down and bleeds because it is not needed to support a baby. Menstruation lasts between three and seven days, and afterwards the lining of the womb starts to build up again. In preparation for a possible pregnancy, the womb lining continues to thicken until day 28.

On about day 14 of the cycle an egg is released from an ovary. This is called **ovulation**. If the egg is not fertilised by a sperm within a few days it dies and is lost from the body with the period.

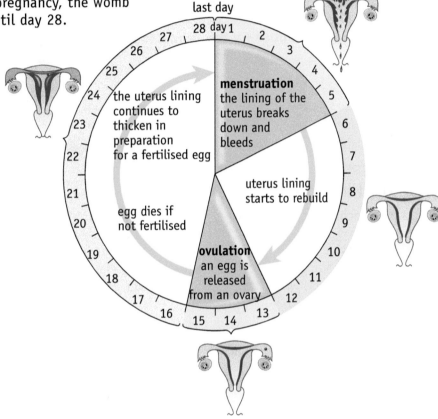

● Figure 3 **The menstrual cycle**

? Questions

1. What is 'puberty'?
2. At what age does puberty happen in girls and boys?
3. Name the hormones that control puberty in boys and girls. Where are these hormones produced?
4. **a)** What is 'ovulation'?
   b) On what day of the menstrual cycle does ovulation usually happen?
5. The lining of the womb thickens and then breaks down every month, unless the woman gets pregnant. Suggest why these changes happen to the womb lining every month.

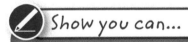

Show you can...

Complete this task to show that you understand puberty and the menstrual cycle.

Draw and complete a table to show the changes that happen in boys and girls during puberty.

For a woman to become pregnant an egg has to be fertilised by a sperm. Usually only one egg is released each month, on about day 14 of the menstrual cycle.

Animals like cats and rabbits do not ovulate in a regular cycle like humans do. Ovulation in these animals is stimulated by sexual intercourse, which means there are many more opportunities for the animal to become pregnant.

● Figure 1 A litter of rabbits. What do you think is meant by the expression 'breeding like rabbits'?

→ Sexual intercourse

When a man becomes sexually excited, his penis swells with blood and becomes hard and erect so he can insert it into the woman's vagina during sexual intercourse. When he **ejaculates**, sperm are released into the vagina. The sperm swim through the cervix and uterus into the oviduct, where they may meet an egg.

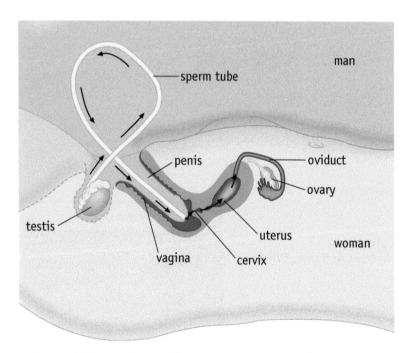

● Figure 2 The positions of the woman's vagina and the man's penis during sexual intercourse. The arrows show the direction the sperm will take to reach an egg

→ Gametes and fertilisation

Gametes are sex cells (the eggs and sperm). Humans and other mammals have **internal fertilisation**. This means the egg is fertilised inside the female's body. Fertilisation occurs when the nucleus of an egg fuses with the nucleus of a sperm.

Human females release one egg each month around day 14 of the menstrual cycle. It enters the oviduct and travels along this tube towards the womb.

If the woman has recently had sexual intercourse, sperm from the male will have swum up from the vagina, through the womb and into the oviduct. The egg may be fertilised if a sperm manages to reach it. Fertilisation usually happens in the oviduct.

Each sperm has a tail so it can swim. They swim in a fluid that contains nutrients. The sperm and this fluid together are called semen. Many sperm are produced to increase the chance that one will fertilise the egg.

Many sperm surround the egg. When one has entered, the outside of the egg changes and no more sperm can get in.

The egg is much larger than a sperm because it contains stored nutrients. This provides food until it implants in the uterus lining.

● Figure 3 Sperm cells are attracted towards an egg, but only one sperm will fertilise it. How many sperm cells do you think a man ejaculates each time he has sex?

→ The embryo

After an egg has been fertilised it travels down the oviduct and divides into a ball of cells, called an **embryo.** It takes about eight days to reach the womb.

When the embryo reaches the womb it sinks into the thickened lining. This is called **implantation**. The embryo develops into a **foetus.** For nine months, the baby develops and grows inside the womb.

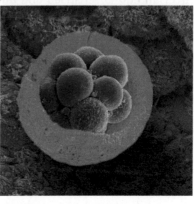

● Figure 4 The fertilised egg has divided into a ball of cells as it moves along the oviduct. How many days old do you think this embryo is?

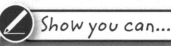

Show you can...

Complete this task to show that you understand sex and reproduction.

Describe the journey of a sperm cell from a testis to an egg inside the woman. Include the names of all the structures it passes through, in the correct order.

? Questions

1 What are gametes?
2 What is 'fertilisation'?
3 Where does fertilisation usually take place?
4 Describe what happens to an egg once it has been fertilised.
5 Many fish release eggs and sperm into the water around them where sperm can swim to the eggs. This is called **external fertilisation** because it happens outside the body.
 Suggest why fewer eggs are produced in internal fertilisation compared to external fertilisation.

During pregnancy, a mother's health is carefully checked by health professionals. The development of the foetus is monitored by ultrasound scans. Doctors get a 3-D picture of what the baby looks like in the womb.

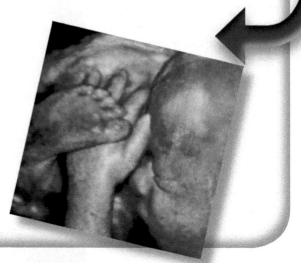

● Figure 1 Why do you think ultrasound scans are used, instead of X-rays, to see the developing foetus?

→ The development of a baby

After an egg is fertilised and until it is eight weeks old it is called an embryo. From then on it is called a foetus.

The embryo implants in the uterus lining and develops blood vessels. An organ called the **placenta** forms, connecting the developing foetus to the mother's bloodstream. The placenta is a very important organ.

The **umbilical cord** is made up of hundreds of blood vessels and joins the foetus to the placenta. The placenta and the umbilical cord carry oxygen and nutrients from the mother's blood to the foetus. Waste products, such as carbon dioxide, pass from the foetus to the mother along the umbilical cord and across the placenta.

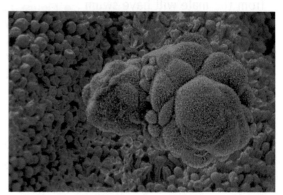

● Figure 2 An embryo. How long do you think it takes before this ball of cells will start to look like a baby?

Pregnancy and health

During pregnancy a woman can help the development of her baby by eating a balanced diet. She will need more energy and protein than usual for the growth of the baby. She should also take sensible exercise. Because of the exchange across the placenta, a pregnant woman should avoid alcohol and smoking. Harmful poisons from both cigarettes and alcohol can pass across the placenta to the foetus. These poisons can lead to problems for the baby. The mother should also avoid other drugs, including some medicines.

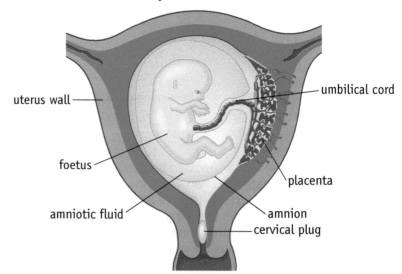

uterus wall

foetus

amniotic fluid

umbilical cord

placenta

amnion

cervical plug

● Figure 3 The developing foetus and placenta

Foetus development

After eight weeks the developing baby is about 4 cm long and is called a foetus. The organs of the foetus are in their final position but are not fully developed. A protective sac called the **amnion** has formed around the foetus. Inside the amnion is **amniotic fluid**. The foetus floats in the amniotic fluid so its weight is supported and it is protected from any bumps.

By five months, the foetus has fully developed lips, eyelids, eyebrows, fingers and toes. The foetus is still growing and its brain still needs to develop.

Most foetuses under five months old cannot survive outside the uterus. Some babies are born a bit early – they are premature. They need to be in an incubator to keep them warm and they may be fed through a tube because they are not yet strong enough to feed on their own.

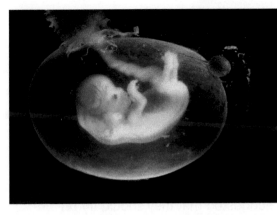

● **Figure 4** A foetus at eight weeks old. How big do you think an eight-week-old foetus is?

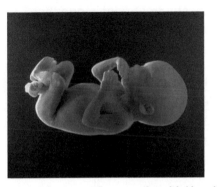

● **Figure 5** A foetus at five months old. How big do you think a five-month-old foetus is?

→ Birth

After 40 weeks the foetus is ready to be born. This length of time is called **gestation**.

Just before birth the womb muscles contract and the cervix dilates (gets wider). The mother uses her abdominal muscles as well as her womb muscles to push the baby down the vagina and out of her body head first. This process is called labour.

The baby starts to breathe air, so the umbilical cord can be clamped and cut. This does not hurt as there are no nerves in the umbilical cord. Soon after the baby is born the placenta is also pushed out.

After a while the baby may take a feed of breast milk, made in the mother's **mammary glands**. The milk has just the right balance of nutrients, and also helps protect the baby from infectious diseases.

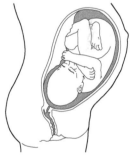

● **Figure 6** The position of the foetus just before birth

? Questions

1 How is the development of a baby monitored during pregnancy?
2 How does a developing foetus get food and oxygen?
3 Suggest how a pregnant woman can protect her own health and her developing baby's health.
4 Describe what happens to the mother and baby during labour and shortly after birth.

✎ Show you can...

Complete this task to show that you understand pregnancy and birth.

Draw a flow diagram to describe the changes that happen to the developing baby during pregnancy.

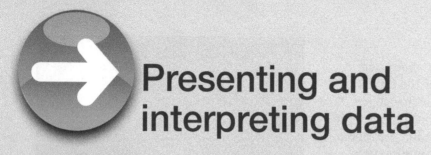

Presenting and interpreting data

➜ IVF clinics

Jenny and Bill have been trying to start a family for several years. Their doctor suggested that they should try **IVF** (in-vitro fertilisation) treatment.

The table below shows the percentage of all IVF treatments that were successful in the UK in 2006. A successful treatment is one that produces a healthy baby.

Age of women given IVF treatment (years)	Percentage success rate
under 35	32
36–37	28
38–39	19
40–42	12
43–44	3
Over 44	3

1 Redraw the data as a bar chart. Put the percentage success rate on the *y*-axis.
2 Jenny is 37 years old. Based on this data what would be the percentage success rate for Jenny?
3 If Jenny cannot get treatment for another year, by how much will the percentage success rate have decreased?

● **Figure 1** Jenny cannot get pregnant naturally. What options might she consider, besides IVF, if she wants a baby?

Jenny and Bill decide to try IVF treatment. The doctor gives them the following information about two local clinics. The table shows percentage success rate and percentage of multiple births (twins and triplets) at the two IVF clinics.

Age of women given IVF treatment (years)	Clinic X		Clinic Y	
	Percentage success rate	Percentage of multiple births	Percentage success rate	Percentage of multiple births
35	29	9	27	13
36	19	8	22	14
37	18	10	19	9
38	15	9	14	14
39	13	10	10	12
40	12	9	9	10

4 Using the data in the table above, which of the two clinics, X or Y, should Jenny and Bill choose? Give two reasons for your answer.
5 They decide they do not want to go to either of these clinics. Use information from this table and the table on page 24 to give a reason why.
6 If Clinic X treated 60 women aged 38 years, how many of these would result in a successful pregnancy?
7 Some people are against the use of IVF treatment. Use information from both tables, and your own knowledge, to suggest why.

As women get older they gradually become less fertile, so less likely to get pregnant. When they are around 50 years of age their ovaries stop releasing eggs and their periods stop. This is called the **menopause**.

IVF treatment now makes it possible for women in their 50s and 60s to have children. Some people think this a good idea, but others are against it. What do you think?

8 Evaluate the use of IVF in older women. Make a list of reasons in favour of older women having a baby, and a list of reasons why older women should not have a baby. Give your answer a conclusion – do you think older women should be given IVF treatment to have a baby?

2.5 Flowers

Many plants rely completely on insects to carry their pollen from one flower to another – this is pollination. If the insects did not do this, the flowers would never produce seeds or fruits. In some extreme cases, only one kind of insect can pollinate one kind of flower.

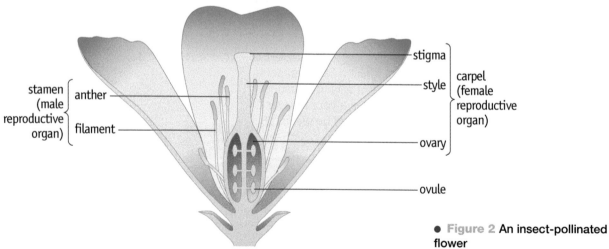

● Figure 1 A honey bee feeding on a flower. How might the decreasing population of bees affect human food production?

All flowering plants have special organs for reproduction – the flowers. Flowers include both male and female reproductive organs. Their final products are seeds. Seeds grow into new plants. This is how plants reproduce sexually.

In order to produce seeds, the flower has to be pollinated. Some flowers are adapted to be pollinated by insects, and others are adapted to be pollinated by wind.

→ Insect-pollinated flowers

Insects are attracted to flowers because of their scent or brightly coloured petals. Many flowers produce a sweet liquid, called **nectar**, which insects feed on.

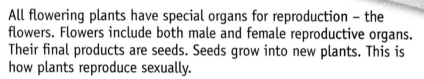

stamen (male reproductive organ) { anther, filament

stigma
style
ovary } carpel (female reproductive organ)

ovule

● Figure 2 An insect-pollinated flower

The female part of the flower is the **carpel**. It is made up of a **stigma**, **style** and an **ovary**. Inside the ovary are **ovules**, each of which contains a female sex cell. The stigma is sticky so that pollen grains stick to it.

The male parts of the flower are the **stamens**. Each consists of a long stalk called the **filament**, which has an **anther** at the top where pollen is made. Each **pollen grain** contains the male sex cell.

When an insect visits the flower to get food, some pollen will stick to its body. The insect then flies off to another flower where some of the pollen may be transferred to the stigma. This transfer of pollen from one flower to another is called **cross-pollination**. If the pollen is transferred to the stigma of the same flower it is called **self-pollination**.

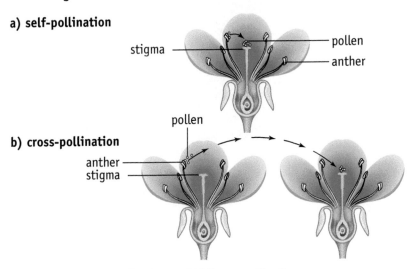

a) self-pollination

stigma — pollen — anther

pollen

b) cross-pollination

anther — stigma

● Figure 3 a) Self-pollination and b) Cross-pollination

Pollen grains are **adapted** for insect pollination. Some pollen grains are hairy or spiky, and others are sticky, so they attach to the insect.

→ Wind-pollinated flowers

Plants like wild grasses and cereal crops are wind pollinated.

Wind-pollinated flowers don't need to be attractive to insects, so they are usually small and do not produce nectar or have large colourful petals. The anthers dangle in the breeze, and the pollen is blown away. The pollen grains are very small and light so they are easily carried on the wind. A lot of pollen is produced, which increases the chances of a pollen grain reaching the stigma in another flower. The stigmas are long and feathery, which gives a large surface area for catching pollen.

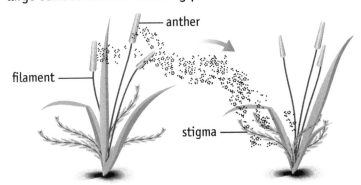

anther

filament

stigma

● Figure 5 A wind-pollinated flower

● Figure 4 Groundsel pollen grains. Why do you think these grains have a spiky surface?

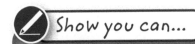

? Questions

1 How are insect-pollinated flowers adapted to attract insects?
2 What is the female part of a flower called? Name all the parts that make it up.
3 What is the male part of a flower called? Name the parts that make it up.
4 Explain how a wind-pollinated flower is adapted for its function.

✎ Show you can...

Complete this task to show that you understand pollination.

Draw a table to compare insect-pollinated and wind-pollinated flowers.

2.6 Seed and fruit formation

Seeds and fruits are an important part of the diet of many animals. Many fruits change colour and start to smell nice as they ripen, and these changes are signals to animals that the fruits are ready to eat.

● Figure 1 Some fruits and seeds that humans eat. Which seeds do you eat that birds or other animals also eat?

→ Fertilisation

A flower is fertilised when the nucleus of a male sex cell, inside a pollen grain, fuses with the nucleus of a female sex cell in the ovule.

The pollen grain lands on the stigma (Figure 2). The sticky fluid on the stigma stimulates the pollen grain to burst open.

A **pollen tube** grows out of the pollen grain and down through the style. The male sex cell moves out of the pollen grain and down the pollen tube (Figure 3).

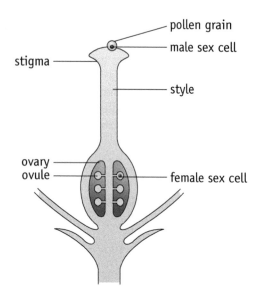

● Figure 2 Pollination

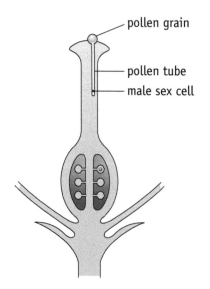

● Figure 3 A pollen tube grows

The pollen tube carrying the male sex cell reaches one of the ovules in the ovary (Figure 4). The tube grows into the ovule and the tip of the tube bursts open so the male sex cell can reach the female sex cell. Fertilisation happens when the two sex cell nuclei join together. The new cell divides to form an embryo.

After fertilisation the petals of the flower shrivel and die. Each fertilised ovule develops into a seed. The ovary around the ovules becomes a fruit.

Fruit or vegetable?

Many foods we call vegetables are actually fruits. For example, courgettes, cucumbers, peppers and pumpkins are all fruits. If you cut them open you will see the seeds inside. Pea pods are fruits, and the peas inside are seeds.

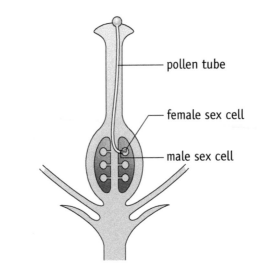

- **Figure 4** Fertilisation happens inside the ovule

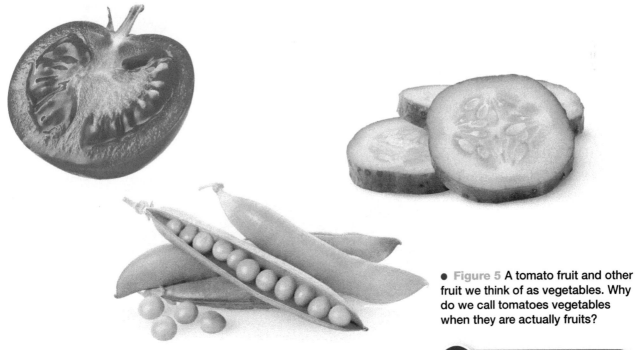

- **Figure 5** A tomato fruit and other fruit we think of as vegetables. Why do we call tomatoes vegetables when they are actually fruits?

Show you can...

Complete this task to show that you understand fertilisation in a plant.

Draw diagrams to show how the nucleus of the male sex cell reaches the nucleus of the female sex cell in order to fertilise it.

? Questions

1 What is 'pollination'?
2 What is 'fertilisation'?
3 Which part of a flower usually forms a fruit after fertilisation?
4 Which parts of a flower form the seeds?
5 After fertilisation the petals on a flower shrivel and die. Why do you think this happens?.

Many designs in nature have stimulated ideas for inventions by humans. Sycamore fruits have wings that make the fruit rotate as it falls from the tree.

● Figure 1 A cluster of sycamore fruits. What modern invention do you think of when you see sycamore fruits spinning down from a tree?

Plants need to spread their seeds away from the parent plant and each other. This is called **seed dispersal**. If the seeds are dispersed far apart there will be less **competition** between them, and therefore more chance of survival. If plants grow too close together they compete for light, nutrients and space. There are several ways that seeds are dispersed.

→ Wind dispersal

Some seeds get blown away from the parent plant by the wind. These seeds are usually very light, or have 'wings' or 'parachutes' to help them travel further. Sycamore fruits, like those in Figure 1, spin as they fall from the tree, and get carried on the breeze.

Poppies have large seed heads filled with small seeds. When the wind blows, the seeds get shaken out of the holes at the top of the seed head.

Dandelions produce many tiny fruits from each flowerhead. Each fruit is attached to a 'parachute' of feathery hairs, which keep the fruit airborne for a long distance. The seed is inside the fruit.

● Figure 2 Poppy seed heads. Can you see the holes that the seeds are shaken out of?

● Figure 3 A dandelion clock. How many fruits do you think there are on this single flowerhead?

→ Animal dispersal

Some seeds are dispersed by animals. Animals eat the soft parts of fruit like cherries or apples, and then drop the seeds on the ground. These fruits are usually brightly coloured and taste nice.

Other seeds get eaten along with the tasty fruit. This happens with blackberries and strawberries. Many birds love to eat blackberries. The seeds pass through the bird's digestive system and are dispersed far from the parent plant in the bird's droppings.

Some fruits have hooks on them. These types of fruit are sometimes called burrs. The hooks catch against an animal's fur, or our clothes, and the fruit gets carried away. The seeds fall out as the animal moves about.

Fruits like acorns are often buried in the ground by animals such as squirrels. The seed is inside the acorn. Squirrels sometimes forget about the buried food store and the acorns germinate into oak trees.

● **Figure 4** Burdock burrs attached to a dog's fur. Why do you think these led to the invention of Velcro?

→ Water dispersal

Coconuts are fruits, and the white flesh and milk inside make up the seed. Coconuts are waterproof and can float. If they fall off a tree into the sea or rivers, they can be carried for thousands of kilometres. If they are washed up on a beach they may **germinate** to form a new tree.

● **Figure 5** A coconut that has started to grow into a new tree. Do you know what a coconut palm looks like?

→ Self dispersal

Plants of the pea family, such as gorse, produce seeds in pods. When the pods dry out they suddenly split open, throwing the seeds out.

● **Figure 6** Gorse seed pods. Why do you think the seed pod splits open when it dries out?

? Questions

1 Name four mechanisms of seed dispersal.
2 How do you think the seeds of a raspberry get dispersed?
3 Give two features of fruits and seeds that are dispersed by the wind.
4 Coconuts are found around the world in tropical and subtropical areas. Explain how they have been dispersed so widely.

✎ Show you can...

Complete this task to show that you understand seed dispersal.

Draw a table naming the four mechanisms of seed dispersal and give at least one plant or fruit as an example for each mechanism.

Planning and designing investigations

→ What affects how fast sycamore fruits fall?

Sycamore fruits are dispersed by the wind.

The fruits spin as they fall. The further from the tree the fruits travel, the better it will be for the new plant's survival. If the fruits fall more slowly they will probably land further from the parent tree.

> 1 Why is it better for a new plant's survival if the seeds are dispersed further from the parent plant?

● **Figure 1** Sycamore fruit containing the seeds

Some students investigated the time it took for model sycamore fruits to fall to the floor. They made their models as shown in Figure 2.

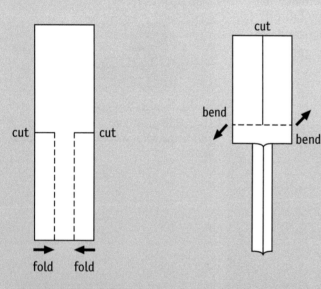

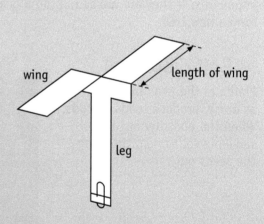

● **Figure 2** How the students made model sycamore fruits

The students decided to investigate the effect of wing length on how long it took for the model to fall to the ground. They started with a model that had wings 10 cm long for the first test, then they reduced the size of the wings for the following tests.

They dropped the models from 2 m above the floor and timed how long each took to fall. They repeated the measurements three times for each model.

> 2 Give one variable that the students controlled in this investigation.
> 3 Give another variable that should be controlled, but would be difficult to keep the same for all the models in this investigation.
> 4 What was the dependent variable in this investigation?

The students' results are shown in the table.

Length of wings (cm)	Time for model to fall 2 m (s)			
	First	Second	Third	Mean
10	2.13	2.19	2.06	2.13
9	1.86	2.03	1.80	1.90
8	1.74	1.78	1.82	1.78
7	1.75	1.38	1.42	1.52
6	1.41	1.26	1.41	1.36
5	1.28	1.29	1.19	

5 What do you think was the greatest source of error in this investigation?

6 How could you alter the students' method to reduce this error?

7 Calculate the mean time for the model with 5 cm wings to fall to the floor.

8 Plot the results in a line graph. Plot the mean time for the model to fall on the *y*-axis and the length of the wings on the *x*-axis. Draw a line of best fit through the points.

9 Use your graph to predict the time a model with 8.5 cm wings would take to fall 2 m.

10 What conclusion can you reach from these results?

11 Can you suggest a reason why the length of the wings affects how quickly the model falls?

12 If a sycamore tree produced a lot of fruit with shorter wings, would this be better or worse for its survival than producing fruit with longer wings? Give a reason for your answer.

The students decided to find out if the mass of the models affected how long it took them to fall to the floor. They used one model with 5 cm wings, but altered its mass by adding different numbers of paperclips to its base.

13 How do you think changing the number of paperclips on the model will affect how quickly it falls? Give a reason for your answer.

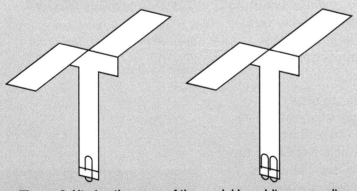

● **Figure 3** Altering the mass of the model by adding paperclips

A habitat is the place where an organism lives. In order for a particular type of organism to survive in its habitat, it has adaptations that are suited to the conditions there. Most organisms would not be able to survive the conditions on a seashore – being exposed to the Sun when the tide is out and being covered in salty water when the tide is in, and having sand and rocks thrown against them by powerful waves.

● **Figure 1** Limpets on a rocky seashore. How do you think limpets are adapted to survive the tides and powerful waves on a seashore?

A **habitat** provides the resources that an organism needs to survive. Animals need food, water and shelter, and plants need space, light, water and minerals. The number of organisms in a habitat depends on the resources available. More light, water and minerals means more, or bigger, plants that can feed more animals.

Different habitats provide different environmental factors such as light, temperature range or water availability, and will therefore be home to different types of organisms. A compost heap and a beech tree are two such habitats that provide different levels of resources.

● **Figure 2** a) A beech tree and b) a compost heap. What types of organism do you think survive in these habitats?

→ Studying habitats

Ecologists study habitats. They measure variables such as temperature, light intensity or moisture. They also use **sampling techniques** to collect data on the types of species living in a habitat. You cannot count all the organisms in a habitat, so random samples are taken. An estimate of **population** size can then be made.

Square frames called **quadrats** are used to estimate plant populations. Ecologists count the number of species in several quadrats placed randomly in the habitat. They can estimate the sizes of populations of various species in the whole habitat. If they carry out these surveys regularly over time, they see changes in populations and can investigate why these changes are happening.

Ecologists study many different types of habitat at the local level (gardens, parks), at the national level (moorland, fenland, chalk downs) and at the global level (ice caps, tropical rainforests).

● **Figure 3** An ecologist sampling plants using a quadrat. Why do you think you couldn't use a quadrat to sample animals?

→ Adaptations to habitats

Polar bears live on the Arctic sea ice. They swim and hunt under the ice or in the sea. They are very well adapted for living in their habitat. Thick white fur insulates and camouflages them. They can walk on ice because the soles of their feet are large and hairy. They are strong animals with sharp claws and teeth to catch seals and walruses. Their keen senses of smell and hearing help them to find their **prey**.

Some organisms can live in a range of habitats. If one habitat disappears they move to another. Foxes are found in both country areas and in towns. Brown rats are able to survive in many habitats, from sewers and fields to cities and coasts.

Other organisms are so well adapted to their habitat that if the habitat gets smaller their numbers fall. Giant pandas mainly eat bamboo and are found only in the mountains of eastern Tibet and southwestern China. Many new roads are being built in China, leaving less space for bamboo to grow. The number of giant pandas is decreasing and they are in danger of becoming **extinct**.

Some animals are adapted to the daily or yearly changes in their habitat. During the winter months in Britain there is less food available. Many small mammals hibernate – they go into a deep sleep and their body temperature drops so they transfer much less energy.

Some birds migrate to a warmer place. Swallows fly to South Africa in the autumn and return to Britain the following spring.

● **Figure 4** A polar bear. Can you think of another adaptation that the polar bear has that helps it survive in the Arctic?

● **Figure 5** A giant panda. What do you think humans could do to stop giant pandas from becoming extinct?

? Questions

1 What is a 'habitat'?
2 List the resources plants and animals need in order to survive.
3 Explain why giant pandas are at risk of becoming extinct.
4 Suggest how foxes are adapted to survive in both country areas and in towns.

✎ Show you can...

Complete this task to show that you understand adaptations.

Describe how a polar bear is adapted to survive in the Arctic.

Some birds spend the summer in the British Isles and then migrate to warmer parts of the world. Swallows arrive in Britain in April and then fly to South Africa in September. Swallows feed on insects as they fly, so the insects could start their lives as part of a food chain in Europe and end their lives as part of a food chain in Africa.

● Figure 1 A swallow. Can you name some animals you think might eat swallows?

→ Food chains

A **food chain** shows the flow of energy and nutrients through organisms in a habitat. The arrows show the direction in which these are transferred.

An example of a food chain in a garden is:

lettuce (producer) → snail (consumer) → thrush (consumer/predator) → cat (consumer/predator)

At the start of a food chain there is usually a plant. Plants use energy from sunlight for photosynthesis to produce food. Plants are called **producers**.

Animals that eat plants or eat plant-eating animals are called **consumers**. Those that eat other animals are also known as **predators**. The animals they eat are **prey**.

Animals that eat only plants are called **herbivores**. Animals that eat only other animals are called **carnivores**. Some animals, like most humans, eat both plants and animals. They are called **omnivores**.

Plants are important because they take in energy from sunlight. The energy in plants passes to the consumers that eat them. This energy then passes to any predator that eats the consumers.

During its lifetime, one snail can eat lots of lettuce, one thrush will eat many snails and a cat will eat more than one thrush. We can show this information in a **pyramid of numbers**.

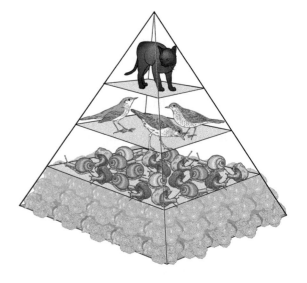

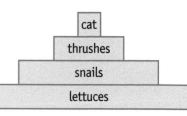

● Figure 2 Pyramids of numbers show the number of organisms at each level in the food chain. The lower diagram is a simple way of showing the proportion of organisms at each level. Not all pyramids are this shape.

→ Ecosystems

An **ecosystem** is the interaction between all the living organisms and all the non-living factors, such as temperature, light, water and shelter, in a habitat. The numbers of organisms in an ecosystem depend on all the living and non-living factors.

Different organisms manage to survive together in an ecosystem because they have different **niches**. Niches are different parts of the habitat or different positions in a food chain. For example, caterpillars and aphids both feed on roses, but they have different niches because caterpillars eat the leaves, whereas aphids suck up sap.

→ Food webs

There are many food chains in any one ecosystem. These different food chains all interact to form a **food web**. Some species eat more than one type of food. For example, a thrush eats earthworms, beetles, caterpillars and snails. We cannot show all this information in a food chain, so we draw a food web like the one in Figure 4.

● Figure 3 Caterpillars eating rose leaves. How many leaves do you think one caterpillar eats in its short lifetime?

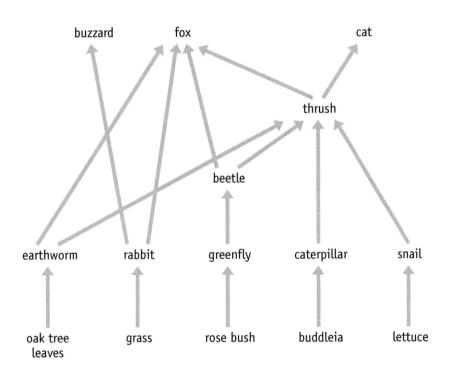

● Figure 4 A food web

If the number of rabbits in this food web decreased the amount of grass would increase, because it would not be eaten as much. However, fewer rabbits would mean less food available for the buzzards and foxes, so their populations would decrease. The buzzard would be affected the most, because the food web shows that rabbits are their only food source in this habitat.

? Questions

1 Explain what a food chain shows.
2 What is a 'producer'?
3 Look at the food web in Figure 4. Draw three different food chains shown in this food web.
4 Look at Figure 4. What would happen to the number of rabbits if the population of foxes decreased? Give a reason for your answer.
5 Describe all the possible effects on the food web if the number of cats decreased.

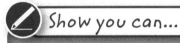

Show you can...

Complete this task to show that you understand food chains.

Draw a food chain and a pyramid of numbers for a habitat where a grass snake eats frogs, which feed on grasshoppers that eat grass.

3.3 Competition and cooperation

There are many examples of humans introducing a new species to an ecosystem in order to control the population of a pest. The Indian mongoose was introduced to Jamaica to control the rat population. However, these mongooses preferred to eat the eggs of local birds and reptiles instead.

● Figure 1 An Indian mongoose. What problems do you think the introduction of the Indian mongoose to Jamaica might have caused?

→ Competition for resources

Resources in a habitat are often limited. This leads to **competition** between organisms of the same species and between members of different species. Plants compete for light, space, minerals and water. Animals compete for food, space and mates. A more successful competitor is more likely to survive.

As the number of individuals in a **population** increases, competition for resources increases. The size of the individuals might also be affected. The cress on the right-hand side of Figure 2 was planted with more seeds in the pot. Many of the seedlings are very small.

● Figure 2 Cress shoots. What do you think the shoots in the right-hand pot are competing for?

Population size

The size of a plant population influences the size of the population of animals that feeds on it. This in turn influences the size of the population of **predators**. The size of the population of predators can also influence the population size of the **prey** organisms.

Any **ecosystem** can only support a certain number of organisms within a population. Once the maximum population size is reached it should stay fairly stable, with some fluctuations. The populations of voles and owls in Figure 3 change but are stable in the long term.

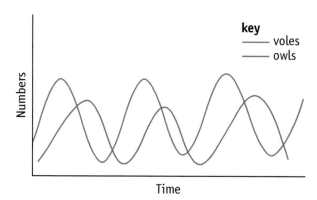

● Figure 3 Graph showing changes in the populations of voles and owls. How can you tell from the graph which organism is the predator and which is the prey?

To maintain the population size, each pair of organisms should produce two surviving offspring. Most organisms produce far more young than this, but many will be eaten by predators or die of natural causes.

Some individuals within a population might be better adapted than others to catch or find food, or escape predation. For example, in a population of reindeer those that have very broad, flat and deeply cleft hooves will be better adapted to run on snow and icy ground. These reindeer will be more likely to escape their wolf predators and survive.

● **Figure 4** Reindeer in the wild. Why will broad, flat and deeply cleft hooves be an advantage on snow and icy ground?

→ Cooperation

Birds of prey, such as eagles, hunt on their own. Those with the keenest eyesight and sharpest claws and beaks will be more successful at capturing prey.

Some animals cooperate with each other in order to obtain food. Wolves prey on animals much larger than themselves, so they work together. They live and hunt in packs. Sometimes half the pack chases the prey while the other half cuts off the prey's escape route. Only one pair of wolves in the pack produce young each year. Fewer young in a pack means that they are more likely to survive. The other females in the pack help to rear these young.

● **Figure 5** A pack of wolves with their prey. How does cooperation help wolves to survive?

? Questions

1 List resources that plants compete for.
2 List resources that animals compete for.
3 Explain what would happen within a population of foxes if there was a shortage of rabbits to eat.
4 Look at Figure 3. Describe and explain the changes in the populations of voles and owls.

✎ Show you can...

Complete this task to show that you understand competition and cooperation.

Make a list of all the factors that can affect the size of a population.

Human effects on the environment

The increasing human population means that we continually use more land to build houses, roads, factories and towns. We also use more land to grow food. More technology means that people are using more energy and raw materials from the Earth. This has led to the destruction of many habitats. More cars and industry also cause more pollution.

● Figure 1 An open-cast mine. How can we reduce the damage we cause to the environment?

→ Food production

Forests and woodland in Britain have been replaced by farmland. This means that food sources and shelter for many animals have been reduced. Trees are important in many **food chains** and they are also **habitats** that house a variety of animals.

The farmland that replaced forests creates new habitats, such as hedgerows between fields. However, modern farming techniques use big machinery that works best in large fields, so many hedges have been removed. This is a physical change that has caused the destruction of a habitat.

During the last 60 years farming has become more productive. Modern farming uses manufactured chemicals such as **fertilisers**, **pesticides** and **herbicides** to increase crop yield. However, these chemicals can significantly affect other animals and plants.

- Herbicides kill weeds and so reduce the food available for some consumers.
- Some weeds are a habitat for small insects, so weed killers can also destroy habitats.
- Fertilisers help plants to grow, but rain can wash fertilisers into rivers. This makes algae grow faster, which blocks light for water plants. The plants die and are decomposed by bacteria that use up the oxygen in the water, so many fish also die.
- Pesticides kill many insects, not just pests, and so reduce food available for birds.
- Some chemicals may remain on the food we eat and so enter the human food chain.
- Some of these chemicals take a long time to break down. When they are taken in by animals, they are stored inside them, so animals higher up the food chain build up a high level of chemicals.

● Figure 2 a) Farmland has replaced woodland. b) Modern farming techniques have led to the removal of hedgerows. How can we produce enough food without destroying more habitats?

→ Bioaccumulation in food chains

DDT was a very successful pesticide. In the 1950s it was sprayed on Clear Lake in California in the USA to kill mosquitoes. Shortly afterwards people noticed large numbers of dead grebes. When the dead birds were analysed they were found to contain very high concentrations of DDT.

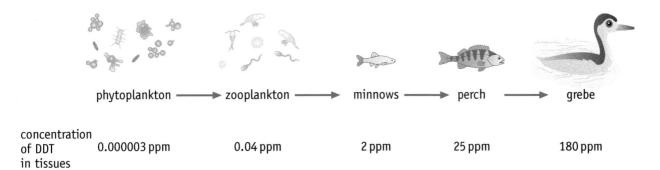

phytoplankton	zooplankton	minnows	perch	grebe
concentration of DDT in tissues				
0.000003 ppm	0.04 ppm	2 ppm	25 ppm	180 ppm

● **Figure 3** A food chain found in Clear Lake showing the concentration of DDT in the organisms

Phytoplankton are microscopic plants. The concentration of DDT in the phytoplankton in Clear Lake was very low, but as the **zooplankton** (microscopic animals) ate many phytoplankton the concentration inside them was much higher. Similarly, minnows eat large numbers of zooplankton, and the DDT gets stored in their fat. Perch eat many minnows and grebes eat many perch in their lifetime, so the concentration of DDT increases along the food chain. This is called **bioaccumulation**.

The concentration of DDT in the grebes killed many birds, or prevented them from reproducing successfully. The use of DDT was banned in the USA in 1972.

? Questions

1 State what fertilisers are used for.
2 Explain why farmers use herbicides and pesticides.
3 Describe three problems caused by the use of chemicals in farming.
4 Look at Figure 3. How many times more concentrated is the DDT in the grebe than in the zooplankton?
5 Explain why organisms at the top of a food chain are affected the most by pesticides such as DDT.

✎ Show you can...

Complete this task to show that you understand how humans have changed the environment in order to produce more food.

Describe how modern farming techniques have maximised food production.

Building scientific awareness

→ Sampling plants on the school field

A class of students was estimating the number of dandelions and daisies on the school field using a **quadrat** that measured 1 m by 1 m. The teacher told the students to drop the quadrat randomly around the field in ten different places.

● **Figure 1** Students counting the number of plants in their quadrat

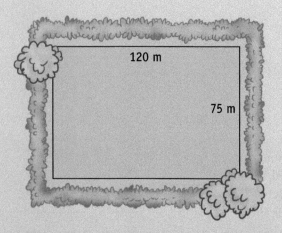

● **Figure 2** The area that the students sampled measured 120 m by 75 m

When they started sampling the students noticed that it was not always easy to count the plants, because some of them were only partly inside the quadrat. Their teacher told them to count any plant that was half way or more into the quadrat, but to ignore anything less than half.

1 What is the area of the quadrat the students used?
2 What is the area of the field that the students sampled?
3 How many of the students' quadrats would fit into the field?
4 Why was it important that the students took random samples?

5 Look at Figure 3. How many daisies should the students count in this quadrat?

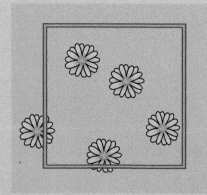

● **Figure 3** When a quadrat is placed at random, only part of the plants at the edge of the quadrat will be inside it

The results obtained by one pair of students are shown in the table.

Quadrat number	Number of daisies	Number of dandelions
1	2	1
2	4	3
3	6	2
4	1	0
5	0	0
6	3	1
7	5	1
8	2	2
9	2	1
10	1	0

6 Calculate the mean number of daisies per quadrat.
7 Calculate the mean number of dandelions per quadrat.
8 Using your answers to Questions 3 and 6, estimate the total number of daisies on the school field.
9 Using your answers to Questions 3 and 7, estimate the total number of dandelions on the school field.

The estimates for the rest of the class are shown in the table below.

Student group	Estimated number of plants on the field	
	Daisies	Dandelions
A	22 300	10 480
B	22 500	9720
C	24 300	10 100
D	9150	23 600
E	23 700	9500
F	23 900	9850

10 a) Which group had some **anomalous results**?
   b) Suggest a reason for these anomalies.
   c) What should you do about these anomalies?
11 Excluding the anomalous results, what is the range of estimates obtained by the students for:
   a) daisies
   b) dandelions?
12 Excluding the anomalous results, do you think the results the class obtained were valid? Give a reason for your answer.
13 Give one conclusion that can be made from the class results in the table above.

The variety of life

In 1938, an unknown fish was caught off the coast of South Africa. A scientist at the local museum identified it as a coelacanth, a fish that was thought to have been extinct for 70 million years.

● **Figure 1** How do you think scientists decide whether they have found a new species, or one that is thought to be extinct?

→ Classifying organisms

Scientists put living organisms into groups, according to their similarities and differences. This is called **classification**. If organisms have similar features, body functions and behaviour, as well as being able to produce fertile offspring, then they are classified as members of the same species. However, members of the same species are not all identical – there is **variation** between individuals.

→ The kingdoms

By observing the characteristics of living things, biologists first classify organisms into one of five groups called kingdoms: animals, bacteria, fungi, plants and protoctists. Protoctists are micro-organisms that scientists still find difficult to classify, so they put them in a group together.

Each kingdom is then divided into smaller and smaller groups, according to how closely biologists think the organisms are related to one another.

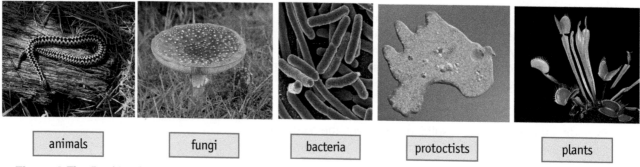

| animals | fungi | bacteria | protoctists | plants |

● **Figure 2** The five kingdoms of living things

→ The animal kingdom

All animals are divided into two large groups – **vertebrates** and **invertebrates**. Vertebrates (such as fish) have a backbone and invertebrates (such as ants) do not.

The five vertebrate groups

Only about 3% of all animals have a backbone.

Fish	Amphibians	Reptiles	Birds	Mammals
• have scaly skin • live in water • swim using fins • breathe using gills • use external fertilisation	• have smooth moist skin • live partly on land and partly in water • have lungs • breed in water • use external fertilisation	• have dry, scaly skin • live in water or on land • have lungs • use internal fertilisation • lay eggs with soft shells	• have feathers and wings • have lungs • keep their bodies warm • use internal fertilisation • lay eggs with hard shells	• have hair or fur • keep their bodies warm • use internal fertilisation • give birth to live young • produce milk

Invertebrate groups

Most animals are invertebrates. Some of the invertebrate groups are shown here.

Molluscs	Arthropods	Echinoderms
• have soft bodies • have a muscular foot • most have a shell • have gills and a rasping tongue • live in water and on land examples: slug, snail, limpet, octopus	• have jointed legs • body is divided into segments • body has a hard outer skeleton • live in water and on land examples: crab, wasp, spider, centipede	• have five matching body parts • have hard, spiny skin • live in water examples: starfish, sea urchin

→ The plant kingdom

Some plants, such as mosses and liverworts, do not produce seeds. Others, such as conifers and flowering plants, do produce seeds.

Mosses and liverworts	Ferns	Conifers	Flowering plants
• small plants without proper roots or stems • no flowers • reproduce using spores • live in damp, shady places	• have roots, stems and leaves • no flowers • reproduce using spores • live in damp places	• have roots, stems and leaves • no flowers • waxy, needle-shaped leaves • reproduce using seeds • live in a variety of places	• have roots, stems and leaves • produce flowers • reproduce using seeds produced inside fruits • live in a wide variety of places

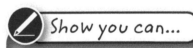

Show you can...

Complete this task to show that you know the five vertebrate groups.

Draw a diagram showing the names and features of the five groups. Include some examples or find pictures to illustrate your diagram.

? Questions

1 Name the five vertebrate groups.
2 What is the difference between a vertebrate and an invertebrate?
3 Name the five kingdoms and give an example of an organism in each kingdom.
4 Explain how scientists classify an organism.

4.2 Variation in living things

One way in which humans vary is whether they are male or female. This is determined at the moment of fertilisation. This is not the case for alligators, crocodiles and turtles. The gender of the offspring depends on the temperature the eggs are incubated at. If they are kept at a temperature of 30 °C they hatch as females. Those incubated at 34 °C hatch as males.

● **Figure 1** A crocodile hatching from its egg. How do you think scientists found out that the incubation temperature affected the gender of the offspring?

→ Variation

You will have noticed that all humans are very similar. At the same time, each of us is unique and so you can recognise people as individuals. This is because we show **variation**.

You can carry out a survey to find out how people vary in a population. When carrying out such a survey you need to collect data systematically. This means you need to plan carefully how to collect and record your data. You also need to ask a large number of people, to give you a large sample. If you only asked a few people, they might all give the same response and you would not see any variation. The bigger the sample, the more likely it is to represent the whole population. A good way to record the data is in a tally chart.

blood group	tally	total
A	ЖЖ ЖЖ ЖЖ III	18
B	ЖЖ I	6
AB	II	2
O	ЖЖ ЖЖ ЖЖ ЖЖ	20

● **Figure 2** A tally chart showing data for blood groups

Discontinuous variation

How are we different from each other? We have different blood groups, gender, eye colour, ear lobes that are either loose or attached, and some of us have disorders such as cystic fibrosis. For each of these differences we can identify a few categories. Each of us has one particular blood group, such as A, B, AB or O. We can either roll our tongue or we cannot. We either have cystic fibrosis or we do not.

These are examples of **discontinuous variation**. There are no in-betweens. We represent discontinuous data using bar charts.

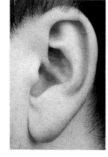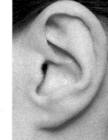

● **Figure 3** Loose and attached ear lobes

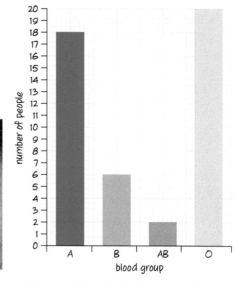

● **Figure 4** Bar chart showing data for blood groups

Continuous variation

Continuous variation is where the data does not fit into groups. For example, height or weight can have any numerical value.

If we measure the heights of a large number of people of the same age, we would get continuous results that can be plotted as a frequency graph or histogram. You cannot plot the height of each person, so you put the heights into groups.

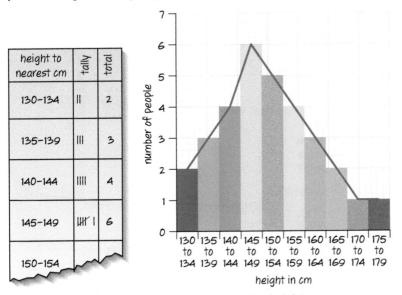

height to nearest cm	tally	total
130–134	II	2
135–139	III	3
140–144	IIII	4
145–149	IHT I	6
150–154		

● Figure 5 A tally chart and frequency graph for height

The people in Figure 6 have a range of heights. Most are in the middle of the range and there are a few at each end of the range. If you only measured a few people, they might be at the ends of the range and not representative of most people of that age.

● Figure 6 People of different heights. What do you think the range of heights for this group is?

? Questions

1 Give two examples of discontinuous variation in humans.
2 Give one example of continuous variation in humans.
3 How should discontinuous and continuous variation be plotted in a graph?
4 Look at Figure 4.
   a) How many people have blood group AB?
   b) Which blood group is the most common?
5 Look at the frequency graph in Figure 5. How many people were included in this survey?

Show you can...

Complete this task to show that you understand how to collect and record data about variation in living things.

Carry out a class survey of two examples of discontinuous variation. Plot your data in two bar charts.

In some families, parents and children have similar strengths in subjects such as engineering, medicine or music.

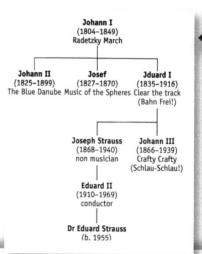

Johann I
(1804–1849)
Radetzky March

Johann II
(1825–1899)
The Blue Danube

Josef
(1827–1870)
Music of the Spheres

Jduard I
(1835–1916)
Clear the track
(Bahn Frei!)

Joseph Strauss
(1868–1940)
non musician

Johann III
(1866–1939)
Crafty Crafty
(Schlau-Schlau!)

Eduard II
(1910–1969)
conductor

Dr Eduard Strauss
(b. 1955)

● Figure 1 The Strauss family tree. How much do you think musical ability is inherited, and how is much due to the environment children are brought up in?

→ Inheritance

You will notice that people who are related to each other usually have some features that are similar. Many children look like one or both of their parents, and some brothers and sisters look very similar. Identical twins are so alike that it is often very hard to tell them apart.

We **inherit** features from our parents. This is controlled by the **genes** that are passed on from parents to offspring. The variation caused by inheriting genes is called **genetic variation**. Genes control our hair colour, whether we can roll our tongue or not, the shape of our nose and many other features.

Offspring are always similar to their parents but never identical to them. This is because at fertilisation a sperm joins with an egg. Half the genetic information for the child comes from the father's sperm and half comes from the mother's egg.

● Figure 2 A family. Which features do these children share with their parents and with each other?

→ Variation in animals and plants

Other animals all show genetic variation. The kittens in Figure 3 are all from the same litter. They have inherited some genes for fur features from their mother and some from their father. They are similar to their parents, but not identical to either of them.

Look at the bull in Figure 5. Farmers use genetic variation to ensure they have animal stocks of a high quality. They will breed animals with useful characteristics, such as good muscle tone. This will mean more meat and a higher price for the animals when they are sold.

Plant breeders also use genetic variation to choose seeds that will produce crops with high yields.

● Figure 3 Kittens with their mother. What do you think controls the colour of the cats' fur?

- **Figure 4** Part of the United Kingdom's Royal Family tree

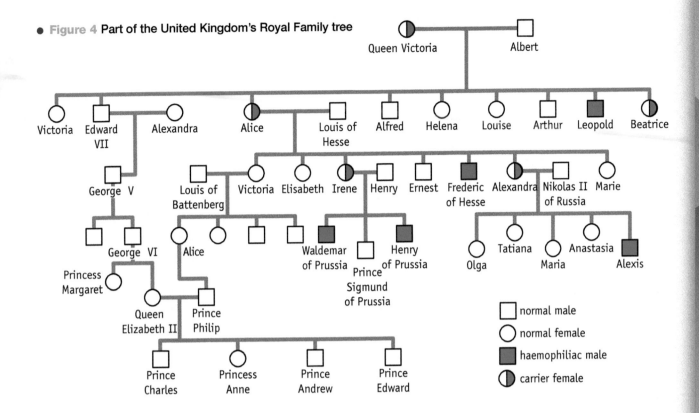

□	normal male
○	normal female
■	haemophiliac male
◐	carrier female

→ Inherited diseases

Some diseases can be inherited. In Figure 4 the family tree shows how **haemophilia** (a disease that increases the time for blood to clot after an injury) has been passed down through the generations of the Royal Family since Queen Victoria.

→ Genetic engineering

Scientists can now transfer genes from one organism to another by a process called **genetic engineering**. Golden rice is a variety of rice that has been genetically modified to produce vitamin A. The rice was developed to be grown in areas where the normal diet is lacking in vitamin A. Eating golden rice could prevent 500 000 cases per year of permanent blindness caused by vitamin A deficiency among children in Africa and Asia.

- **Figure 5** A prize-winning bull. What characteristics do you think the farmer looked for to produce this bull?

? Questions

1. Give one feature you have that is similar to your father, and one that is similar to your mother.
2. State what controls the features that we inherit from our parents.
3. **a)** Look at the Royal Family tree in Figure 4. Suggest why Queen Victoria did not have haemophilia.
   b) Explain why none of Queen Elizabeth II's children have the gene for haemophilia.

Show you can...

Complete this task to show that you understand the cause of genetic variation.

Explain why children have similar features to their parents but are not identical to either of them.

4.4 Environmental variation

The twins in the photograph are genetically identical, and yet there is still variation between the two sisters. For example, their fingerprints will be similar, but differences between them will be visible.

● Figure 1 Identical twin sisters. What similarities and variations can you see between these sisters?

Not all variation is due to inheritance. Variation caused by environmental factors is called **environmental variation**. For example, someone might have a scar, false or missing teeth, a suntan or dyed hair. These are not caused by their genes, but by environmental factors.

→ Environmental variation in animals

Piglets from the same litter have the same parents but there are often size differences between them. This is not solely a genetic variation. The womb was their environment before they were born. Some piglets may have received more food and grown faster when they were in the womb environment.

→ Environmental variation in plants

Plants also show environmental variation. Both crops shown in Figure 2 were grown from seeds from the same parent plants. The variation in height is due to the conditions in which the plants were grown. The taller, stronger plants got more minerals from the soil and this produced healthier growth.

● Figure 2 Variation in plants. What do you think caused the differences in height and colour of these crops?

The colour of the flowers on mophead hydrangeas is affected by the pH of the soil they are grown in. In acidic soil the flowers are blue, in acid–neutral soil the flowers are mauve and in alkaline soil they are pink.

→ Disease and diet

Disease and diet are environmental factors that can cause variation in humans. The child in Figure 4 is very small for her age because she has been severely undernourished (she has not had enough food). People with tuberculosis (TB) are often shorter than average, because the disease affects lung function and may reduce oxygen supply to tissues. Many people who have had chickenpox are left with small scars where the heads of the spots were knocked off.

● Figure 3 **Hydrangea flowers. Do you think seeds from these flowers would always produce offspring with the same flower colours? Why?**

→ Variation in learning

Some children are much better than others at maths or at learning a foreign language. They may have inherited the potential to develop these skills from their parents but environmental factors, such as diet and education, allow some of them to develop those skills more. A good diet helps your brain to develop. Intellectual ability is an example of continuous variation and is the result of genetic and environmental variation working together.

All children inherit the ability to learn language. However, they need to be brought up in a learning environment where parents talk to them. This usually needs to happen by the time the child is two years old. This is when the brain is most receptive to learning language. Feral children, who have been brought up by animals such as monkeys or wolves, show behaviour like that of the animals that brought them up. They often have few or no human language skills.

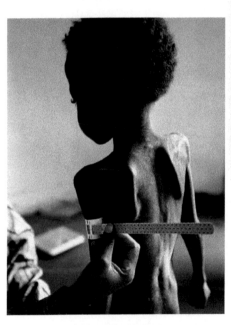

● Figure 4 **An undernourished child. What other effects do you think a poor diet can have on children?**

? Questions

1 Give two examples of environmental variation that you or members of your family have.
2 If plants are grown in poor soil, how might this affect the way they grow?
3 If a mophead hydrangea is grown in alkaline soil, which colour flowers will it probably have?
4 Disease can be an environmental factor that affects variation. List some examples of how disease might cause variation in people.

✎ Show you can...

Complete this task to show that you know some environmental causes of variation.

Make a list of some examples of environmental variation in humans, and say what might have caused these.

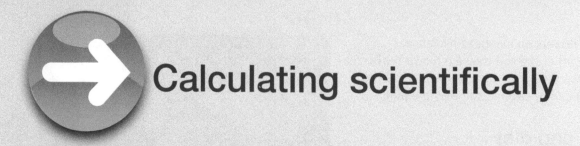

Calculating scientifically

→ Population data and investigating variation

Population data

A census is a way of collecting and recording information about the people in a particular **population**. The first census of the United Kingdom took place in 1801, prompted by fears that as the population was growing very quickly there could be food shortages. Censuses have taken place in the United Kingdom every ten years, except during World War II. The most recent census was carried out in 2011.

● **Figure 1** Who makes up the population of the United Kingdom?

1 Why do you think it is useful to collect data about the population of a country?

From the 2011 census data, the population of North Yorkshire was 598 376. Of these people 294 110 were males.

2 Is gender an example of **genetic** or **environmental variation**, or both?
3 Is gender a **continuous** or **discontinuous** example of variation?
4 What percentage of the population in North Yorkshire in 2011 were males?
5 a) How many people in North Yorkshire in 2011 were females?
  b) Round your answer to the nearest thousand people.
6 Round the number of males in North Yorkshire in 2011 to the nearest thousand people.
7 Plot your answers to Questions 5b and 6 in a graph.

Investigating variation

Some students were asked to investigate the variation in the length of privet leaves on a hedge. Each student measured the length of 20 leaves. One student's results are shown in the table.

Length of privet leaves (mm)			
43	21	19	51
21	41	49	30
40	39	23	50
12	30	16	42
29	39	45	37

● **Figure 2** Privet leaves. Why are these good for investigating variation?

8 Is leaf length an example of genetic or environmental variation, or both? Give a reason for your answer.
9 Is leaf length a continuous or discontinuous example of variation?
10 What is the range of leaf length the student measured?
11 What is the mean leaf length the student measured?
12 Before this data can be plotted in a graph it needs to be organised into groups. Draw a table similar to the one below and record the data. You will need to decide what leaf lengths you will put into each group. The first group has been done for you.

Length range for group (mm)	Tally	Total
10–19		

13 Use the data in your table to draw a graph.
14 Make a list of factors that might affect the length of privet leaves on a hedge.

5.1 The importance of plants

Many square kilometres of forest have been removed in Britain. Most of the cleared land is used for farming, and the rest is used for housing and roads.

● **Figure 1** How do you think the removal of forests has affected oxygen production?

→ Plants and photosynthesis

Plants and **algae** (simple plants) take in carbon dioxide from the air and water from the soil to produce oxygen and food (in the form of glucose). Plants do this by absorbing energy from sunlight. The process is called **photosynthesis**, and it can be represented by an equation:

$$\text{water} + \text{carbon dioxide} \xrightarrow[\text{chlorophyll}]{\text{light}} \text{glucose} + \text{oxygen}$$

→ Plants as producers

Plants are called producers because they make food, which provides energy and **biomass** for other organisms. Nearly all food chains begin with a plant. Without plants we would not have any food.

We would not have any meat, fish, eggs or milk to eat without plants, because almost all animals obtain their food and energy from plants.

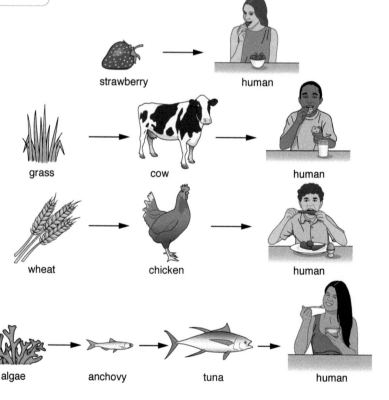

● **Figure 2** Some food chains that include humans

54

The process of photosynthesis also helps to maintain the correct balance of carbon dioxide and oxygen in the air. Forests are particularly valuable for maintaining the oxygen content of the air. When humans remove large areas of rainforest to build roads and houses, the oxygen produced is greatly reduced. The habitats of other organisms are also destroyed.

There is much concern that the concentration of carbon dioxide in the atmosphere is increasing due to the burning of fuels and **deforestation**. High levels of carbon dioxide contribute to **global warming**.

● Figure 3 Deforestation in Borneo. Why are large areas of tropical rainforest being cut down and burnt?

→ What do plants give us?

Without plants we could not survive. We rely completely on plants to produce the oxygen we need to respire, and to supply food. Plants are also important in the water cycle.

Plants provide us with many other products. Various plant materials are used for building, such as trees for timber and cereal plants for straw to thatch roofs. Many different plants provide fibres that we use to make cloth, ropes and paper. We also obtain many chemicals from plants, including medicines, and use plants to make **biodegradable plastics**. We use wood as a fuel, and we also grow oilseed rape, oil palms and sugarcane to produce fuels.

→ Chemosynthesis

There are areas in deep oceans where it is so dark that plants cannot survive there. Scientists were therefore very surprised to find huge communities of different animals living there. The organisms cluster around very hot, acidic springs called **hydrothermal vents**.

In the hydrothermal vents bacteria absorb chemicals from the hot water and use them to produce carbohydrates for the rest of the food chain. They do this by a process called **chemosynthesis**. The bacteria are eaten by animals such as vent limpets, which are eaten by other animals such as vent crabs.

● Figure 4 A hydrothermal vent at the bottom of the ocean. What could be the producer for food chains near hydrothermal vents?

? Questions

1 Name the two products of photosynthesis that make plants essential for the survival of animals.
2 Give four other uses we have for plants.
3 Draw a food chain for organisms living near a hydrothermal vent. Include the producer for this food chain.

✎ Show you can...

Complete this task to show that you understand the importance of plants.

Draw a spider diagram with plants in the middle to show why plants are important to us.

5.2 Photosynthesis

Commercial fruit and vegetable growers use greenhouses to create the best conditions for photosynthesis and so get good plant growth.

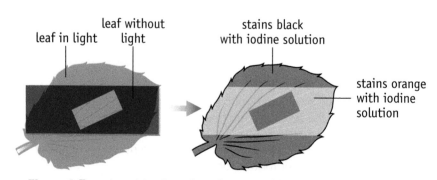

● **Figure 1** Sweet peppers growing in a greenhouse. How do you think the grower maintains the best conditions for photosynthesis in the greenhouse?

Plants make food by **photosynthesis**. Plants look green because the cells contain a green substance called **chlorophyll**. Chlorophyll absorbs energy from sunlight, which plants transfer into chemical energy during photosynthesis.

The raw materials for photosynthesis are water and carbon dioxide. The products are glucose and oxygen. This is shown in the equation:

$$\text{water + carbon dioxide} \xrightarrow[\text{chlorophyll}]{\text{light}} \text{glucose + oxygen}$$

Some of the glucose is used in **respiration** and the rest is turned into chemicals such as **starch**, **fats**, **proteins** and **DNA**.

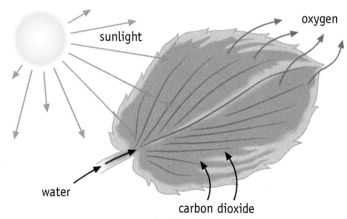

● **Figure 2** How a leaf obtains the raw materials for photosynthesis

→ Photosynthesis experiments

Experiments can be done to see if light, carbon dioxide and chlorophyll are needed for photosynthesis, by testing leaves to see if they contain starch. If they do, the leaf has been photosynthesising and has stored some of the glucose as starch. Iodine solution is used to test for starch. It goes black if starch is present and stays orange if starch is not present.

● **Figure 3** Experiment to show that plants need light for photosynthesis

Is light needed for photosynthesis?

Cover part of a leaf with card and leave the plant in bright light for a few days. The covered part of the leaf will not get light. If the leaf is then tested for starch you would see the results shown in Figure 3. The part of the leaf that was covered does not make starch, but the parts of the leaf that were exposed to light do make starch.

Is carbon dioxide needed for photosynthesis?

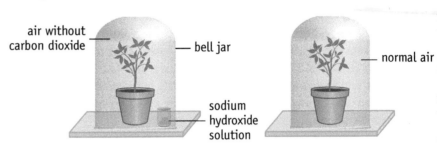

1 Put one plant in a bell jar with a beaker containing sodium hydroxide solution. The sodium hydroxide solution absorbs carbon dioxide from the air.

2 Put a control plant in another bell jar, but without sodium hydroxide solution, so this plant will have carbon dioxide in the air around it.

● Figure 4 **Experiment to show that plants need carbon dioxide for photosynthesis**

3 Put both plants in bright light for a few days, then remove a leaf from each plant and test them both for starch.

The leaf from the plant without carbon dioxide would not contain any starch, but the leaf from the plant in air would contain starch.

Is chlorophyll needed for photosynthesis?

Some plants have leaves that are not green all over. They are called **variegated leaves**. The green parts contain chlorophyll but the white parts of the leaf do not. If a plant with variegated leaves is put in the light for several days and then a leaf is removed and tested for starch you would see the results shown in Figure 5.

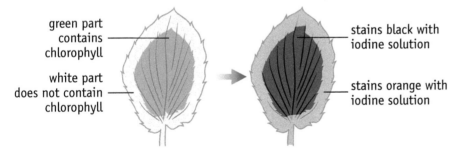

● Figure 5 **Experiment to show that plants need chlorophyll for photosynthesis. Only the green part of the leaf makes starch**

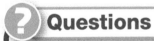 **Questions**

1 Describe what chlorophyll is, and its function in plants.
2 Write down the word equation for photosynthesis.
3 Describe what happens to the glucose formed in photosynthesis.
4 Describe the test for starch.
5 Look at Figure 5. Give a conclusion you can reach from these results.

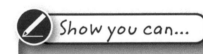 Show you can...

Complete this task to show that you understand photosynthesis.

Make a drawing of a leaf and show on it the raw materials and products of photosynthesis.

5.3 Leaf structure

The weeping fig originally came from areas in Asia and Australia where there is a lot of rain. The leaves are shiny, curve downwards and have a pointed tip called a drip tip. Botanists (scientists who study plants) think this helps the rain to wash straight off the leaf, preventing any microscopic plants from settling on the leaf and blocking its sunlight.

a drip tip

● **Figure 1** The leaves of a weeping fig. How many other plants can you find that have a drip tip?

→ How are leaves adapted for photosynthesis?

Leaves are organs that are specialised for photosynthesis.

Leaves are usually flat and broad so that they have a large surface area to absorb light and carbon dioxide. They are very thin so that there is a short distance for carbon dioxide and water vapour to travel to reach the photosynthesising cells. Water is transported from the roots to the leaves through the **midrib** and veins. Glucose made during photosynthesis is transported away from the leaf to the rest of the plant in the veins.

Some plants can turn their leaves to follow the Sun, so that they absorb as much light as possible during the day. Other plants position their leaves so that they do not overlap or shade each other.

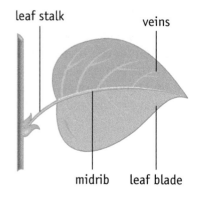

leaf stalk

veins

midrib   leaf blade

● **Figure 2 The structure of a leaf**

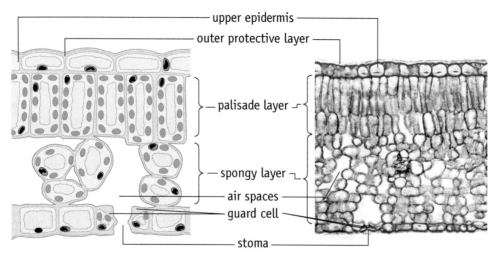

upper epidermis

outer protective layer

palisade layer

spongy layer

air spaces

guard cell

stoma

● **Figure 3** The internal structure of a leaf. Where do you think most photosynthesis occurs in the leaf?

→ The internal structure of a leaf

Look at the magnified cross-section of a leaf in Figure 3. The **upper epidermis** is a transparent layer that light passes through to the layers below. The cells do not have chloroplasts, so they do not photosynthesise. The cells are tightly packed together to protect the leaf.

The **palisade layer** is the main photosynthetic layer. The cells are packed with chloroplasts, containing chlorophyll, so they can absorb as much light as possible.

Cells in the **spongy layer** also have some chloroplasts to absorb any light that passes through the palisade layer. There are large air spaces between the cells so gases can easily pass to and from the palisade cells.

The lower epidermis protects the underside of the leaf. The only cells of the epidermis that have chloroplasts and can photosynthesise are the **guard cells**. These are pairs of specialised cells that have a hole called a **stoma** between them. The guard cells control the size of the stoma. Carbon dioxide moves into the leaf and oxygen moves out of the leaf through the stomata (plural of stoma).

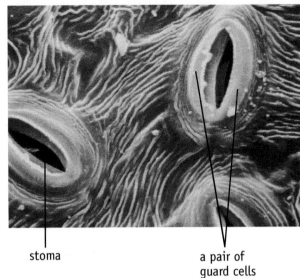

stoma

a pair of guard cells

● Figure 4 Guard cells and stomata. How do you think the stomata open and close?

→ Experiment to show that plants make oxygen

If pond weed is put in bright light it will photosynthesise and make a gas that can be collected as shown in Figure 5. The gas escapes from the leaves through the stomata.

When a full tube of gas is collected it can be tested. If a glowing splint is put into the tube, the splint relights, showing that the gas is oxygen.

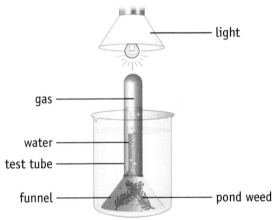

light

gas

water

test tube

funnel

pond weed

● Figure 5 Experiment to show that plants make oxygen

? Questions

1 Why do leaves look green?
2 In which two layers of a leaf does photosynthesis happen?
3 Describe what stomata are, and describe their function.

5.4 Mineral salts and fertilisers

Hydroponics is a method of growing plants in a solution of mineral salts. No soil is needed. The plants are grown in large greenhouses.

● **Figure 1** Lettuces grown using hydroponics. What do you think are the advantages of growing lettuces using hydroponics rather than soil?

→ Mineral salts

Plants photosynthesise to make glucose, but they would not survive long without small amounts of chemicals called **mineral salts**. Plants absorb mineral salts from the soil through their roots. The mineral salts are dissolved in the water in the soil.

→ Fertilisers

Gardeners and farmers add **fertilisers** to the soil to help their flowers or crops to grow. Fertilisers contain the mineral salts that plants need. Manure is a natural fertiliser, and there are chemical fertilisers like those shown in Figure 2b.

● **Figure 2** a) A farmer spreading manure on his field and b) some chemical fertilisers. What do you think NPK stands for?

→ Essential elements

Mineral salts contain the chemical elements that are essential for healthy plant growth. For example magnesium is needed to make chlorophyll. Without enough magnesium the leaves look yellow instead of green.

NPK fertilisers contain three other important elements – nitrogen, phosphorus and potassium. Figure 4 shows what these elements are needed for in a plant.

plant lacking magnesium

plant after fertiliser has been added

 Figure 3 Magnesium is needed to make the green pigment chlorophyll

Element in mineral salt	nitrogen (N) provided by nitrates	phosphorus (P) provided by phosphates	potassium (K) provided by potash
What it is needed for	to make proteins that are needed for growth	for healthy roots	for production of flowers and fruits
Effects of a shortage	• stunted growth • pale coloured leaves • weak stems	• short roots and stems • small purple leaves • low yield of fruit	• brown edged leaves • poor fruit production • poor resistance to disease

● **Figure 4 The effects of a shortage of mineral salts on plant growth**

Questions

1 Why do plants need magnesium?
2 Name the elements that NPK fertilisers contain.
3 A gardener noticed that her plants were not growing well. They were very small with weak stems. Name the element you think these plants were lacking.
4 Describe what hydroponics is.
5 Suggest some advantages of growing lettuces using hydroponics rather than growing them outside in soil.

Show you can...

Complete this task to show that you understand why plants need mineral salts.

Draw a table to summarise the effects of different elements on plant growth.

61

Presenting and interpreting data

→ Fertilisers and crop growth

Crops that are harvested every year remove large amounts of **mineral salts** from the soil. Scientists calculated how much nitrogen, phosphorus and potassium a tonne of wheat removed from the soil in one year.

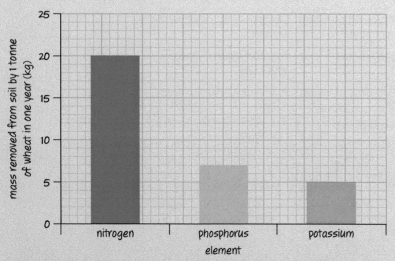

● **Figure 1** Bar chart to show the mass of elements removed from the soil in a year by a tonne of wheat

1 What mass of phosphorus did 1 tonne of wheat remove from the soil in one year?
2 How many times more nitrogen than potassium was absorbed by the wheat?
3 Use the data in the bar chart to suggest why farmers add **fertilisers** to fields where they grow crops.

The table shows the yield of wheat when different fertilisers are used on fields.

Fertiliser used on field	Wheat yield in tonnes per field per year
none	95
nitrogen	159
nitrogen and phosphorus	167
nitrogen, potassium and phosphorus	211

4 Plot the data from the table as a bar chart.
5 In order for the results for each element to be compared, a controlled test should have been done. Using the units given in the table for the yield, give one factor that should have been controlled.
6 a) Work out how much each element (nitrogen, phosphorus and potassium) increased the yield of wheat.
   b) What conclusion can you reach from your answer to Question 6a)?

Other factors as well as mineral salts can affect plant growth. These include water availability, carbon dioxide concentration, light intensity and temperature.

Scientists can investigate the effect of one of these variables on plant growth. They grow two large batches of the same type of plant in a controlled environment, and only change one variable at a time so that they can investigate it properly.

Plant growth can be measured by measuring either the height or the mass of plants. The growth of seedlings at different temperatures was measured over 13 days. Two batches of seedlings were grown, one batch at 10 °C and the other at 20 °C. Ten seedlings were randomly taken from both batches every other day and their heights were measured. The mean heights were then calculated. The results are shown in the table.

● **Figure 2** A scientist investigating plant growth. Who might benefit from this kind of research?

Day	Mean height (mm)	
	Grown at 10 °C	Grown at 20 °C
1		10
3	15	24
5	30	40
7	35	55
9	40	69
11	50	75
13	55	90

7 Why were the seedlings selected randomly?
8 The mean height of the seedlings grown at 10 °C on day 1 is missing from the table. The heights of the ten plants sampled on day 1 were: 8.4, 7.2, 8.6, 7.5, 7.9, 8.1, 8.3, 8.0, 8.4 and 7.6. Calculate the mean height for these ten seedlings.

Figure 3 shows the data for the seedlings grown at 20 °C plotted as a line graph.

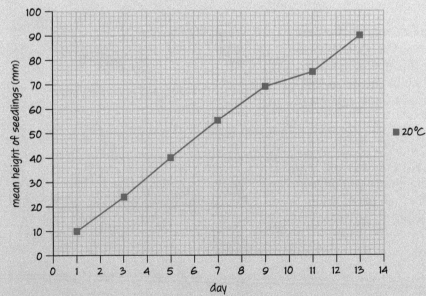

● **Figure 3** A graph to show the growth of seedlings at 20 °C

9 Make a copy of the graph and plot the results for seedlings grown at 10 °C. Join the points with straight lines.
10 Why should the lines start from day 1 and not be extrapolated back to 0?
11 a) What conclusion can you reach from these data?
   b) Suggest an explanation for your conclusion.

New research is being carried out to find the effects of different nutrients on different genes. Some nutrients affect genes in different ways. They might switch on genes for diseases like Alzheimer's disease. Related research is trying to identify the gene that makes some people more likely to become obese.

● Figure 1 A scientist carrying out human gene sequencing. How might this research be used to provide people with personalised diet plans?

→ The seven food groups in a balanced diet

In order to stay healthy we need to eat a **balanced diet**. This means eating the correct balance of the seven food types – **carbohydrates**, **proteins**, **fats**, **vitamins** and **minerals** as well as **fibre** and water.

We also need to eat foods that give us the right amount of energy. If we take in too much energy in our food we may become overweight or even obese. If we take in too little energy we will lose weight and may become ill.

Carbohydrates

Carbohydrates are the main source of energy in the diet. Starches and sugars are carbohydrates. Energy is released from carbohydrates in **respiration**. This energy is used to keep us warm, for movement, to build chemicals and to make new cells. The brain also uses a large amount of energy.

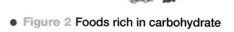

● Figure 2 Foods rich in carbohydrate

Proteins

Proteins are needed for growth and repair of cells and tissues. Enzymes are proteins, and cell membranes contain protein. Babies and children need more protein than most adults because they are growing. Pregnant women also need more protein than usual in their diet.

Fats

Fats are needed as a store of energy. They contain almost twice as much energy per gram as carbohydrates or proteins. Fats also form an important part of cell membranes and nerve cells.

● Figure 3 Foods rich in protein

Vitamins

Vitamins are substances that the body needs in small amounts. There are many different vitamins that are needed for different functions. Vitamin A is needed for healthy eyes and skin. Vitamins in the B group are needed for healthy nerves, respiration and making blood cells. Vitamin C is needed for healthy joints and blood vessels. It is needed to make connective tissue, which supports and binds together other body tissues. Lack of vitamin C causes a disease called **scurvy**. Vitamin D helps us to absorb calcium and keep our bones and teeth strong. Sunlight causes a fat under the skin to change into vitamin D. Lack of vitamin D causes a disease called **rickets**. The bones in the legs are too weak and so they bend.

● Figure 4 **Foods rich in vitamin C**

Minerals

Minerals are elements that are needed in small amounts for different functions. Iron is needed to make red blood cells. A shortage of iron causes **anaemia**, which makes a person feel very tired and lacking in energy. The best source of iron is red meat. Calcium is needed for healthy teeth and bones.

Fibre

Fibre is made of cellulose, the chemical that forms cell walls in plants. Humans cannot digest fibre, but it is important in helping the movement of food through the digestive system. Too little fibre in the diet can cause constipation.

Water

Water makes up about two-thirds of our bodies. The chemical reactions in cells take place in solution, and water is needed to transport substances around the body in the blood. We lose water in urine and sweat and this must be replaced. Without water a human would die in a few days. You should try to drink just over a litre of water a day.

● Figure 5 **Water is essential for life**

? Questions

1 Suggest what would happen if you ate too much high-energy food.
2 Which food type is needed for growth and repair of cells?
3 Why is it important that we eat fats and oils?
4 The diseases caused by a lack of a vitamin or a mineral are called **deficiency diseases**. Name the deficiency disease caused by a shortage of each of the following: vitamin C, vitamin D, iron.
5 Suggest what might happen if we did not eat enough carbohydrates.

Show you can...

Complete this task to show that you know why you should eat a balanced diet.

List the seven food types and explain what each is needed for in the body.

6.2 Food tests

Food tests are carried out in food-testing laboratories. Food scientists can find the amounts of different food types and the amount of energy in a food, or test its safety. They can also find out if the food being tested is what is stated on the label.

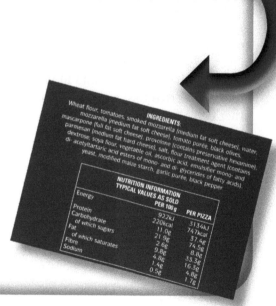

● **Figure 1** A food label for a pizza. Why is it useful to put the ingredients and nutritional value of food on labels?

You can carry out food tests in school to see which food types different foods contain. Remember to wear eye protection.

→ Testing for starch

Method
Add a few drops of **iodine solution** to your food sample.

Result
If **starch** is present the sample will turn black. If starch is not present the sample will stay orange.

● **Figure 2** Testing foods using iodine solution. Does the food being tested here contain starch?

→ Testing for glucose

Method
Add some water to your food in a test tube to make a solution. Add an equal volume of **Benedict's solution** and heat the test tube in a water bath.

Result
If **glucose** is present the solution will turn green, orange or brick-red, depending on how much glucose is present. If no glucose is present the solution will stay blue.

● **Figure 3** Testing foods using Benedict's solution. Which tube contains glucose?

→ Testing for protein

Method
Add some water to your food in a test tube and shake it thoroughly. Then add a few drops of **Biuret reagent**.

Result
If **protein** is present the solution will turn lilac or purple, depending on how much protein is present. If no protein is present the solution will stay blue.

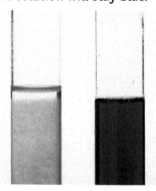

● **Figure 4** Testing foods using Biuret reagent. Which tube contains protein?

→ Testing for fats

Method

Mix the food with a third of a tube of ethanol and shake well. Filter the mixture to remove any solids. Add some clean water to the filtrate.

Result

If **fat** is present a white emulsion will form and float on the water. If no fat is present the solution will stay colourless and clear.

● **Figure 5** Testing foods using ethanol and water. What does this result show?

→ Measuring the energy given out by burning foods

You can also investigate the amount of energy in different foods by burning them and using the heat given off to heat up a measured volume of water. If you weigh out the same mass of different foods you can compare the energy content of them. The food that warms the water up the most must have released the most energy. Remember to wear eye protection.

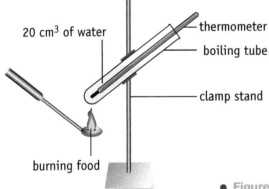

● **Figure 6** Investigating the energy content of foods

? Questions

1 Name the solution you would use to test to see if a food contained starch.
2 Name the type of food Biuret reagent is used to test for.
3 A food solution was heated with an equal volume of Benedict's solution. After heating, the solution was blue. What conclusion can you reach from this result?
4 a) A student burnt a crisp that weighed 0.5 g. The temperature of a tube of water increased by 25 °C. Calculate the temperature rise per gram of crisp.
  b) The student then burnt a dried pea that weighed the same as the crisp. Predict whether the temperature rise of the water would be the same, more than or less than that for the crisp. Suggest a reason for your prediction.

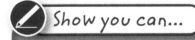

Show you can...

Complete this task to show that you know how to carry out food tests.

Describe all the steps involved in testing a food for starch and glucose. Describe the results you might see.

The digestive system

If a patient has problems digesting or absorbing foods a doctor can use an endoscope to see inside the stomach and digestive system. The endoscope is put down the patient's throat and photographs or samples of tissue can be taken to identify the cause of the problem.

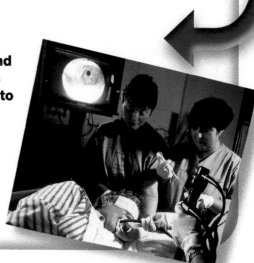

● Figure 1 An **endoscope** being used to see inside a patient's stomach. Why is using an endoscope better than carrying out an operation?

→ Digestion

The functions of the **digestive system** are to digest and absorb foods. **Digestion** is the breakdown of large, insoluble food molecules into smaller, soluble ones. These can then be absorbed into the bloodstream.

Food is physically broken down by chewing food in the mouth, using the teeth. The pieces of food are then small enough to be swallowed. More physical digestion happens when food is churned in the stomach.

Chemicals called **enzymes** are added to the food in the mouth, stomach and small intestine.

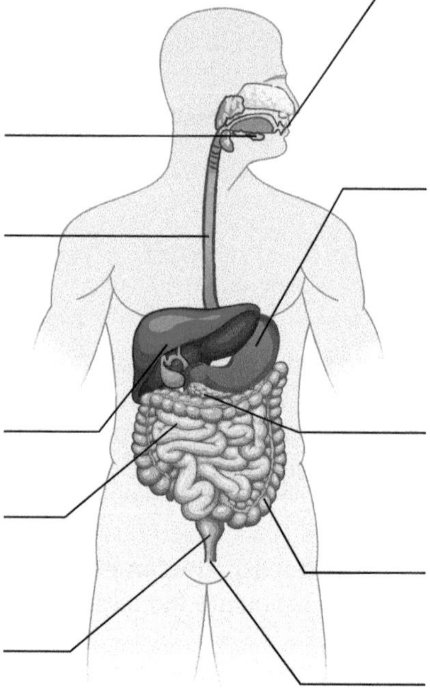

salivary glands

oesophagus (gullet)
a muscular tube that pushes food into the stomach

liver

small intestine
digestion is completed here and the products are absorbed into the blood

rectum
undigested foods (faeces) are stored here

mouth
food is chewed and saliva is added from the salivary glands. The saliva contains enzymes

stomach
a muscular bag that churns food with digestive juices. These juices contain hydrochloric acid to kill bacteria, and enzymes to digest food. Food may stay in the stomach for several hours

pancreas
produces enzymes that are released into the small intestine

large intestine
water is absorbed into the blood

anus
faeces are passed out of the body when you go to the toilet

● Figure 2 The human digestive system

Enzymes are biological **catalysts** that speed up chemical reactions. The enzymes in the digestive system speed up the breakdown of starch, proteins and fats, which are too large to be absorbed. The products of this chemical digestion are small enough to be absorbed through the wall of the small intestine into the blood.

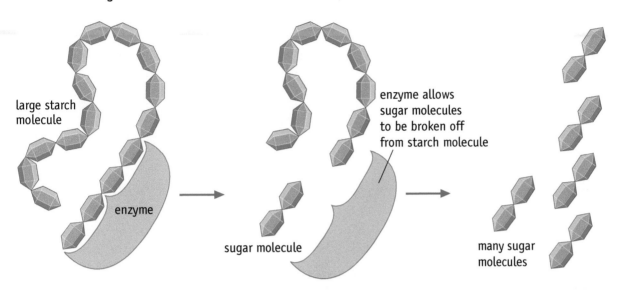

large starch molecule

enzyme

enzyme allows sugar molecules to be broken off from starch molecule

sugar molecule

many sugar molecules

● Figure 3 An enzyme breaks down a large starch molecule into small sugar molecules

→ Absorption

The small intestine is adapted for efficient **absorption**. It is very long and its wall has many finger-like projections called **villi**. The villi increase the surface area for absorption. The wall of a villus (singular of 'villi') is only one cell thick so food substances can easily cross into the bloodstream. Inside each villus is a good blood supply to pick up the food molecules and transport them around the body to be used or stored.

→ Gut flora

Gut flora are bacteria that live in the digestive system. Many of these live in the large intestine where their presence reduces the growth of harmful micro-organisms. They are also useful because they can digest some chemicals that humans cannot digest, and because they produce some vitamins that are absorbed into the blood.

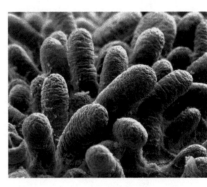

● Figure 4 The small intestine is lined with villi. How do you think these increase the total surface area of the small intestine?

? Questions

1 Explain how food is physically broken down in the digestive system.
2 What is an enzyme?
3 List all the organs in the digestive system that produce digestive enzymes.
4 Describe what happens in the large intestine.
5 Explain the benefits of the bacteria that live in our digestive system.

✎ Show you can...

Complete this task to show that you understand the digestive system.

Make a list of ways in which the small intestine is adapted for efficient absorption of food molecules.

Presenting and interpreting data

→ Do people all need the same amount of energy?

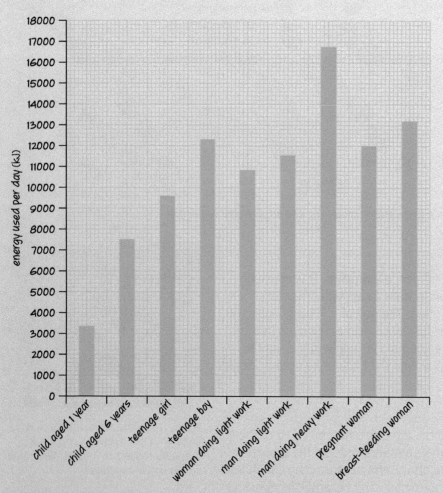

● **Figure 1** The amount of energy used per day by different people

1 How much energy does a teenage girl use per day?
2 How much more energy does a teenage boy use per day?
3 Use information in the bar chart to list four factors that affect the amount of energy a person uses per day.
4 a) Suggest what would happen to a person who took in more energy in their food than the amount of energy they used every day, and this continued for a few months.
  b) Suggest how this might affect their health.

→ Calculating the amount of energy in different foods

A student calculated the amount of energy in different foods by burning them and using the energy released to heat water in a boiling tube. The apparatus she used is shown in Figure 2.

The student burnt 1 g of dry food each time, and measured the same volume of water into the boiling tube. When the food had completely burnt she measured the temperature rise of the water. Her results are shown in the table.

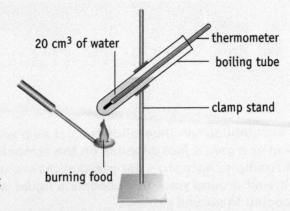

● **Figure 2** Apparatus to measure the energy given out by burning foods

Food	Temperature rise (°C)
apple	2
peanut	60
toast	15

Her teacher said she could calculate the amount of energy that had transferred to the water using the equation:

energy transferred to water (J) = mass of water (g) × 4.2 × temperature rise (°C)

She used 20 cm³ of water each time, which is 20 g, so:

amount of energy transferred to water from apple:
= 20 × 4.2 × 2
= 168 J

5 Calculate the amount of energy in joules (J) transferred to the water from the peanut and the toast.
6 a) Which food released the most energy?
  b) Suggest a reason for this.
7 When tested in a laboratory, the actual amount of energy in 1 g of each food is:
  • apple: 1960 J/g
  • peanut 24500 J/g
  • toast 10600 J/g.
  Suggest why the values from the student's experiment are much lower than these values.

Food scientists can accurately calculate the amount of energy in different foods using a bomb calorimeter. Study Figure 3 carefully before answering the following questions:

8 A measured amount of food is put into the steel bomb and set on fire. How do you think the food is set on fire?
9 Where will the energy from the burning food transfer to?
10 What is the purpose of the stirrer?
11 Give two ways in which this piece of apparatus is more accurate than the method the student used in Figure 2.

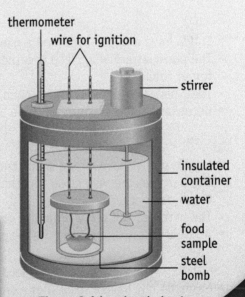

● **Figure 3** A bomb calorimeter

Particle model

All substances can theoretically exist as a solid, a liquid or a gas; it just depends on the temperature. For example, nitrogen – the gas that makes up 70% of the air around you – will become a liquid when it is cooled to around –200 °C.

● **Figure 1** A steaming bowl of food. What solids, liquids and gases can you see around you every day?

Scientists believe that everything is made from very tiny particles. They use this idea, called **particle theory**, to explain what happens to solids, liquids and gases when they are heated and cooled.

Particles are not static, they all have some energy due to their movement. One of the key differences between the states of matter (solid, liquid or gas) is how quickly the particles are moving, and this links to how close the particles are to each other.

● **Figure 2** Particles in a solid

→ The three states of matter

Solids

- Solids have a fixed shape and a fixed size.
- The particles are very close together and held in place by strong forces (bonds).
- Their particles cannot move around, but they do vibrate.
- Because the particles cannot move around, a solid has a fixed shape.

● **Figure 3** Particles in a liquid

Liquids

- Liquids do not have a fixed shape but they do have a fixed volume.
- The particles are very close together. Most of the particles touch each other.
- The particles can move around.
- A liquid can flow and take the shape of its container.

Gases

- Gases do not have a fixed shape or a fixed volume.
- The particles move around all the time and spread out. This is why a gas fills its container.
- A gas can be compressed into a very small space – this pushes the particles closer together.

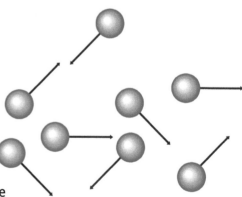

● **Figure 4** Particles in a gas

→ Explaining expansion

As a solid is heated, its particles gain more energy. This makes the particles vibrate faster and they move further apart. The solid expands.

The particles themselves don't expand – their size stays the same. Their mass stays the same too.

 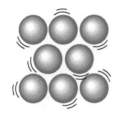

● **Figure 5** When particles are heated they gain additional energy because of increased movement. This leads firstly to **expansion** and then to a change of state

Gases and liquids also expand when they are heated, because their particles gain more energy and move around more.

Expansion can be a problem – for example, railway lines can expand and buckle in hot weather.

Expansion can be useful too – thermometers work because the liquid inside expands when it is heated and contracts when it is cooled.

? Questions

1 Describe the differences in how particles move in each of the three states of matter.
2 What is the difference between having a fixed shape and having a fixed volume?
3 When engineers design metal bridges they do not make them in one piece. The joints they include often include overlaps of metal, which allow some movement. Explain why they are designed this way.
4 Explain why gases can easily be compressed but solids and liquids are very difficult to compress.
5 Explain, using diagrams, why an air-filled balloon goes down over the course of a week, even if the neck is securely tied.

 ## Show you can...

Complete this task to show that you understand the differences between the three states of matter.

Consider these everyday examples, which can be difficult to categorise as solids, liquids or gases.

· ice cream · foam · toothpaste · sand

For each one describe whether they have a fixed shape and whether they can flow. Then choose which state of matter matches them best and explain your choice.

1.2 Changing state

Understanding changes of state is important when working with materials. Glass blowing, casting metal and cooking are all activities in which changes of state are used to create a finished product.

● **Figure 1** Mercury is the only metal that is a liquid at room temperature. Why do you think it is a good choice to use in a thermometer?

A material changes its state because energy is either added or taken away from its **thermal store**.

Adding energy to a solid makes its particles vibrate further and faster until the bonds between particles weaken enough for the particles to move past each other – the solid has now melted into a liquid.

Adding energy to a liquid makes its particles move around further and faster until they have enough energy to break their bonds and become free to move anywhere. When this happens the liquid has become a gas.

A substance can become a gas in three ways: **evaporation**, **boiling** and **sublimation**.

● **Figure 2** A summary of state changes

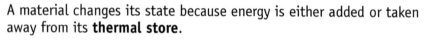

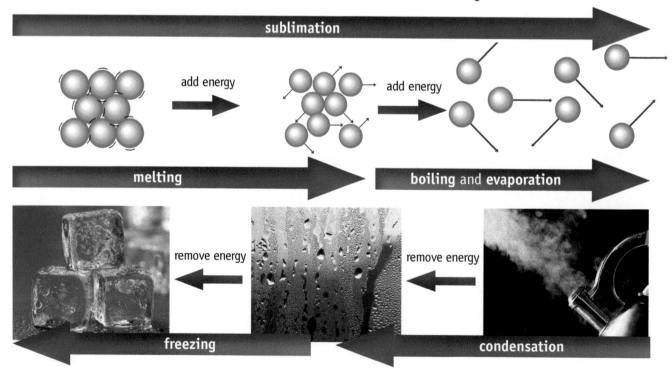

→ Evaporation

- Only the particles at the surface of the liquid are involved – they are the only particles with enough energy to break free.
- The liquid doesn't bubble.
- Evaporation happens at all temperatures.

→ Boiling

- All the particles in the liquid are involved – they all have enough energy to break free.
- The liquid bubbles.
- Boiling happens at a fixed temperature, called the boiling point.

→ Sublimation

Some substances have a melting point and a boiling point so close together that when they are heated they do not melt but turn straight into a gas. This is called sublimation, and iodine is an example of such a substance.

→ Melting

Melting, like boiling happens at a fixed temperature for a given liquid. These temperatures are different for different liquids.

Liquid	Melting point	Boiling point
Pure water	0 °C	100 °C
Mercury	–39 °C	360 °C

→ Removing energy

Change of state is a physical change – it can be reversed.

Cooling a gas takes energy from its particles so they slow down and move closer together. The gas **condenses** into a liquid.

Cooling a liquid takes energy from its particles so they can't move around and can only vibrate. The liquid **freezes** into a solid.

When they change state, particles do not change size or mass – they just arrange themselves differently. If you melt 1 kg of solid iron you will get 1 kg of liquid iron!

? Questions

1. Explain the difference between the arrows on the liquid particles and those on the gas particles in Figure 2.
2. Which arrangement has the most energy: solid or gas?
3. Why is mercury a good choice of liquid to use in a thermometer? What problems could there be if water was used instead?
4. Compare the effect of heating a liquid and heating a gas.
5. Puddles are an example of water evaporating. Describe this process and explain how it differs from boiling. Explain why we don't see it happening.

✎ Show you can...

Complete this task to show that you understand changes of state.

In dry weather washed clothes can be hung outside to dry. For wet washing to dry, the liquid water particles need to be changed to gas particles and be moved away from the washing. Explain, using diagrams, why the washing dries quicker on a warm and windy day

1.3 Diffusion and gas pressure

When substances are in the gas arrangement they have some unique properties. Scientists and engineers capitalise on these properties to make gases useful. Hydraulic machines use the compressibility of gases to allow heavy loads to be moved with minimal effort.

● **Figure 1** How can understanding how gases act under pressure influence modern sports equipment?

→ Diffusion

Moving particles will spread out if they have space to move into. This is called **diffusion**. Diffusion happens even if there is no breeze or current.

During diffusion, particles move from areas where they are strongly concentrated to areas where there are fewer of them. This means that the particles are spread out more evenly after diffusion than before.

Gases diffuse easily because their particles can move freely.

Liquids can also diffuse. Diffusion in a liquid usually happens slowly.

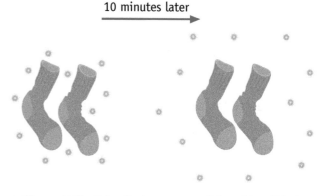

10 minutes later

● **Figure 2** Particles that cause smell travel by diffusion

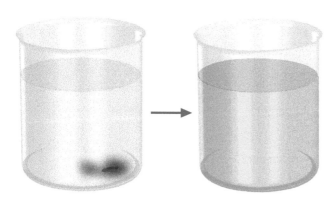

● **Figure 3** Using a coloured dye can allow us to observe diffusion in liquids

How quickly particles diffuse depends on a number of factors. For example:

● the size of the particles – the smaller and lighter the particles are, the faster they diffuse
● the temperature – the higher the temperature of the gas or liquid, the more energy the particles will have and the faster they will diffuse
● the difference in concentration between where the particles were when they started and where they are moving into – the greater the difference, the faster they will diffuse
● any other particles that are in the way – they could bump into other particles, which would slow the diffusing particles down.

→ Gas pressure

As gas particles move around they hit objects. This causes **gas pressure**.

The faster the gas particles move, the more force they exert so the greater the pressure is.

The hotter the gas, the more energy its particles have and the harder they hit the container.

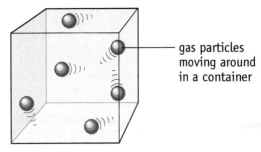

● **Figure 4** As the gas particles hit the sides of the container they cause gas pressure

gas particles moving around in a container

→ Air pressure

Air particles moving around cause air pressure (**atmospheric pressure**). Air pressure acts in all directions.

If air pressure is greater on one side of a surface, the unbalanced forces can make the object collapse. If you remove the air from inside the can shown here the can will collaps.

Aerosols, steam turbines, drinking straws and pneumatic drills all use air or gas pressure to work.

Two thousand years ago, the Egyptian scientist, Hero of Alexandria, designed machines that used air pressure to move objects.

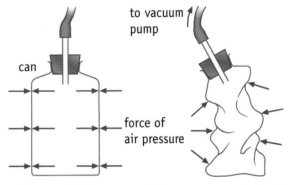

● **Figure 5** As the gas particles are pumped out of the can the sides collapse

to vacuum pump

can

force of air pressure

? Questions

1 Why is diffusion usually slower in a liquid than in a gas?
2 When smell particles diffuse across a room, what are the other particles that could get in the way?
3 Draw a series of three particle diagrams to show the stages of diffusion that relate to Figure 3.
4 Explain what would happen to an inflated balloon if you put it into the freezer.
5 Explain how the air in a bicycle tyre causes pressure and describe in detail what will happen to the gas pressure on a very warm day.

Show you can...

Complete this task to show that you understand diffusion.

You are going to investigate the effect of temperature on the rate of diffusion.

a) Describe the equipment you will need for this investigation and why.
b) Explain the factors that you will have to keep the same in each part of the investigation.

1.4 Introducing density

Not all solids are the same. In some solids the particles are packed so tightly together that you get a lot of mass in a small volume. These very dense materials are useful if you want strength and hardness, such as in the iron used in the head of a sledgehammer.

● **Figure 1** *RMS Titanic*. How can a ship made of steel, and with a mass of over 40 000 tonnes, float?

Density measures how concentrated the mass of an object is.

- If an object has its mass spread over a large volume, its density will be low.
- If all the mass is concentrated in a small space, its density will be high.

→ Properties of solids

1 Solids have a fixed volume and a fixed shape.
2 However, they can be **elastic**. This means they return exactly to their original shape after bending or stretching.
3 Solids can be hard. This means they will not easily change shape when a force is applied.
4 Solids can also be strong, meaning a large force is needed to break them. This is because their particles are close together and held in place by strong bonds. It can take a lot of force to break these bonds.
5 Solids can be dense. This is because their particles are packed closely together. This means that there is a large mass in a small volume of material.

● **Figure 2** Can you think of any other tools or toys that use elasticity to work?

6 Some solids can have lower density. A good example is aluminium, which is used to make aircraft. The advantage of this metal is that it has the hardness and strength of a solid but a lower mass, which reduces the force needed to lift off.

● **Figure 3** Other than helping the aircraft lift off, what other advantages are there to keeping its mass to a minimum?

→ Properties of liquids and gases

Liquids have a fixed volume but no fixed shape.

Gases have no fixed volume and no fixed shape.

Gases expand to fill the space available. They can also be compressed into a very small space.

→ Density and buoyancy

An object will float if its density is less than the density of the liquid it is in. Some types of wood are less dense than water, so they will float on water. Most metals are denser than water, so they won't float on water.

Materials that float easily are described as 'buoyant'.

As a general rule, solids are denser than liquids, and liquids are denser than gases.

Solids – highest density, because their particles are closest together
Liquids
Gases – lowest density, because their particles are furthest apart

However, there are some exceptions.

Liquids with a low density can float on liquids with higher density – oil floats on water. Gases with a low density will float on high-density gases too – warm air will rise and float on colder air.

Water is an example of a substance in which the solid arrangement is less dense than the liquid. This is why ice will float on water.

● **Figure 4** Liquids take the shape of their container

● **Figure 5** What could happen if the exit for the steam was blocked?

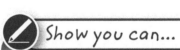

● **Figure 6** What could the advantage be to the living organisms in aquatic environments once there is a layer of ice floating on the surface of the water?

? Questions

1 Why are solids harder than liquids?
2 What would it mean if a solid was described as strong, hard and dense?
3 Why is it an advantage to run cars on liquid fuels, such as petrol, rather than solid fuels, such as coal or wood?
4 Explain why the density of a metal object decreases when the object expands.
5 An object, such as a rubber ring, can be made to float on water by filling it with air. Explain, using particle diagrams, why this occurs.

✎ Show you can...

Complete this task to show that you understand density.

Choose two items from your school bag or pencil case. How could you determine which one has the greater density? Think about what equipment you might need and what results you might expect.

Asking questions and making predictions

→ Applying the particle model

Models help us understand new ideas by relating them to things we already know about. We use the particle model of solids, liquids and gases to explain their behaviour.

We can also use models to make predictions about how similar materials will behave. This allows us to develop scientific questions about real-life observations.

Hardness and scratch resistance

Solids can be hard because their particles are held together by strong bonds. The stronger the bond between the particles in a material, the harder the material is.

A common way to compare the hardness of materials is to rub two objects against each other and observe which one becomes scratched. The softer of the two objects will show either temporary or permanent changes to its shape. Try this now – holding a piece of paper in the air, rub your pencil gently against it.

1 Which is harder? How do you know?

Now let's make a prediction.

2 What do you think will happen if you rub a metal pen tip against the wood of your pencil? Explain your prediction in terms of the hardness of the materials you are comparing. Then test it out. What did you observe?

Diamond is one of the hardest natural materials. It can only be cut using small pieces of other diamonds.

3 How would you break the diamond apart when it is so hard?

Elasticity

Solids can be elastic because the forces (bonds) between their particles resist the stretching force trying to pull the particles apart.

Some solids are more elastic than others – because the forces holding the particles together are different.

James was investigating the elastic properties of a rubber band. You could repeat his investigation if you have a rubber band available and make your own observations.

● **Figure 1** Why would you expect some dust to be created during a scratch test?

particles held together with stretchy bonds

when you pull on the material, the bonds stretch

the bonds spring the particles back to their original positions

● **Figure 2** The stretchy bonds between particles in an elastic material allow it to store energy

The first thing James wanted to do was observe what happened when the rubber band stretched. He started by sketching out how the particles are arranged in the unstretched rubber band.

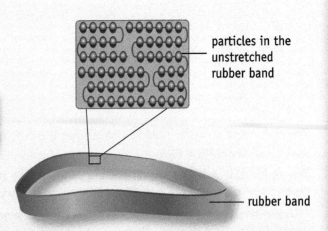

particles in the unstretched rubber band

rubber band

4 Predict what will happen to the thickness of the rubber band as it is pulled tight. Explain your prediction and draw a particle diagram to show the arrangement in the stretched rubber band.

The next thing James was interested in was whether the rubber band would always spring back to its original length, however far it was stretched.

5 Do you think stretchy bonds have a limit? Give an example of something you have observed to support your idea.
6 What might happen to the stretchy bonds between the particles if the rubber band is stretched very far?

● **Figure 3** The arrangement of particles in an unstretched rubber band

Everyday particles

Mary was going on a school trip. While she was having her packed lunch she made some unusual observations. She wanted to try and explain them using her scientific knowledge and understanding.

Mary was feeling rather thirsty, so she decided to drink one of the cartons of juice that she had brought. She pushed the straw into the carton and drank all of the juice. She noticed that the carton had become squashed. She thought that while the squashed carton was left on the side it would expand again and regain its original shape, but it didn't.

7 Why do you think this happened?
8 How could Mary get the carton to return to its original shape?

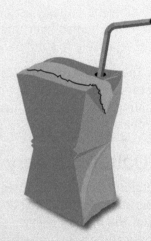

The day of the school trip was a hot one, and Mary had filled a plastic bottle with water but had left it on the coach during the day. When she got back to her seat she saw that the lid had popped open and some of the water had spilt.

● **Figure 4** A squashed drinks carton

9 Why do you think this happened?

2.1 Atoms

The Ancient Greeks had ideas about elements and atoms 2000 years ago.

Some thinkers believed that everything was made up from four elements – Air, Fire, Water and Earth.

Others thought that everything contained small particles called atoms. The word 'atom' actually comes from a Greek word, which means indivisible ... so they had the idea that an atom could not be divided into anything smaller.

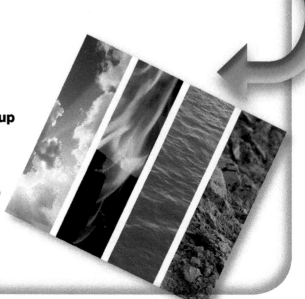

● **Figure 1** Air, fire, water, earth. Which chemicals that you know do you think could be put into these ancient groups?

We are surrounded by millions of different chemical substances. To help us make sense of these, scientists put them into groups, such as metals, rocks, plastics or smart materials. All of these substances are made up of 90 different naturally occurring building blocks, which we call **atoms**. Even in the human body, blood, muscle, bone and brains are made out of relatively few different atoms combined in different ways.

It is the combination of atoms in a substance that determines it properties, including its colour, hardness, melting point and density.

→ Dalton's theory

After John Dalton developed his theory about atoms in the early 1800s, they became part of the modern chemist's ideas. Dalton's theory can be summarised:

- everything is made up of tiny particles called atoms
- atoms cannot be destroyed
- atoms cannot be broken up into smaller pieces
- in an **element** all the atoms are identical
- atoms can join together and make bigger particles (now called molecules)
- different types of atoms can combine to make **compounds**
- atoms combine in simple whole number ratios when they form compounds.

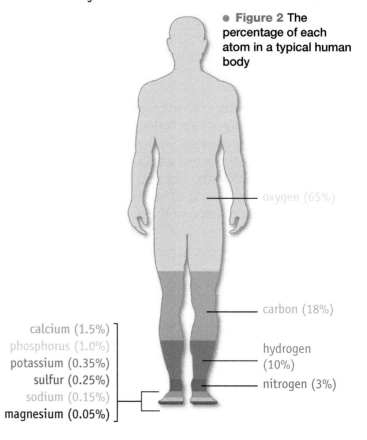

● **Figure 2** The percentage of each atom in a typical human body

oxygen (65%)

carbon (18%)

hydrogen (10%)

nitrogen (3%)

calcium (1.5%)
phosphorus (1.0%)
potassium (0.35%)
sulfur (0.25%)
sodium (0.15%)
magnesium (0.05%)

An atom is the smallest part of a substance that has the properties of that substance. It is the basic particle from which all substances are built up. The desk, the air, and even you are made up of atoms.

Atoms are far too small to see, even with a microscope. It can be really difficult to picture the size of a single atom. Imagine the point of a very sharp needle; this contains approximately one million atoms in that tiny space.

→ Arrangements of atoms

The 90 naturally occurring types of atoms that make up different substances are different from each other. For example:

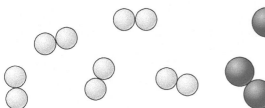 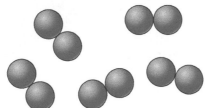 

- hydrogen gas contains only hydrogen atoms

- oxygen gas contains only oxygen atoms
- hydrogen atoms are different from oxygen atoms

- gold atoms are different from both of these. This is why gold is a heavy metallic solid and hydrogen is a light gas

● **Figure 3** Different arrangements of atoms give rise to different chemical substances

Scientists have believed for over 200 years that different substances are made out of different atoms, because the model works well. This theory explains, for example:

- why some new materials can be made by combining other materials
- how some chemical substances can be split into other simpler substances
- why substances are different from each other.

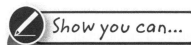

Show you can...

Complete this task to show that you understand about atoms.

Draw some particle diagrams for the following substances based on the information given. Include a key for the different atoms you have used.

- Water is a compound made up of one oxygen atom in the middle with a hydrogen atom each side.
- Carbon dioxide is made up of a central carbon atom with two oxygen atoms, one on each side.
- Methane, the gas used for Bunsen burners, is arranged with a central carbon atom surrounded by four hydrogen atoms.

? Questions

1 How could we show the differences between atoms when we draw particle diagrams?
2 Oxygen, carbon and hydrogen are the three most common atoms in the human body. What percentage do they make up all together?
3 Compare and contrast the terms 'atom', 'element', 'molecule' and 'compound'.
4 Look at Figure 3. How would you tell from the images that these are all examples of elements?
5 Explain how it is possible to have so many different materials from relatively few different types of atom.

2.2 Atomic structure

When scientists suggested the modern theory of the structure of the atom in 1932, it allowed the explanation of many properties of elements to be concluded – from the colours, masses and melting points of the elements to predicting the range of chemical reactions that they take part in.

● **Figure 1** How do the planets move in our solar system in relation to the Sun?

→ Atomic theories

John Dalton was the first scientist to expand on the ancient Greeks' ideas about atoms. As they did, he also concluded that everything is made up of tiny particles called atoms and that these atoms cannot be broken up into smaller pieces.

A century later his ideas were to be questioned. In 1897 a scientist called J J Thomson discovered the electron, a tiny negatively charged particle that was much smaller than an atom. He concluded that as these particles were so small they could have only come from inside atoms and called it a **subatomic particle**.

As all atoms were neutral, that is they had no charge, Thomson proposed that these tiny negative electrons must be stuck in a cloud of positive charge. His 'plum-pudding' model was so called as it resembled the currants spread through a Christmas pudding.

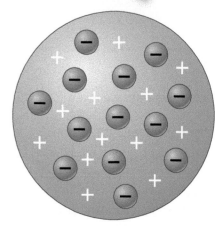

● **Figure 2** Thompson's plum-pudding model of the atom

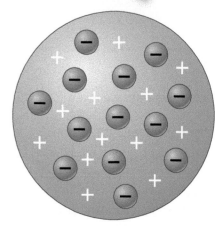

orbital electrons, negatively charged

nucleus, containing positively charged protons

● **Figure 3** Rutherford's atomic model of 1911

Just over 10 years later, a scientist called Ernest Rutherford and two of his students were carrying out some further investigations and found new evidence that did not fit with Thomson's model. Bringing together all their ideas with those from previous scientists, they came up with what is now accepted as the basic model for atomic structure.

The atom was shown to be mostly empty space. A tiny positive nucleus, made up of subatomic particles called **protons**, was in the centre with the even smaller negative **electrons** orbiting around the outside.

→ Subatomic particles

Finally, in 1932, a third subatomic particle was discovered: the **neutron**, a particle with the same mass as a proton but having neutral charge.

The subatomic particles can be summarised as follows:

Name of subatomic particle	Relative mass	Relative charge
proton	1	+1
neutron	1	0
electron	$\frac{1}{1800}$ (almost nothing)	−1

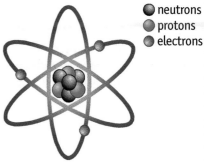

● **Figure 4 The modern-day model for atomic structure**

Once scientists had this model for atoms they could begin to explain several observations.

The number of subatomic particles in the nucleus determined the mass of the atom – the atomic mass, or **relative atomic mass**. Scientists could also begin to explain how different elements behaved. The attraction of the positive nucleus for the negative electrons began to explain why some elements are solids at room temperature while others are gases.

When atoms combined, it was the electrons that were either swapped or shared, therefore, the orbiting electrons accounted for the chemical properties of the atoms, not only affecting whether they interacted with other atoms and formed bigger particles, but also how quickly it happened.

The Periodic Table had been developed in 1871, grouping elements together. Now scientists had a way to explain why these groups behaved in similar ways. The elements were related to other members of their group by the number of electrons in the outermost orbitals of the atoms.

? Questions

1 Which two subatomic particles are in the nucleus of an atom?
2 In which order were the three subatomic particles discovered?
3 Why would it be the outermost electrons that explained the grouping of elements?
4 Why are the numbers of protons and electrons the same for a particular atom?
5 Explain why the number of protons and neutrons affects physical properties whereas the number of electrons affects the chemical properties of an element.

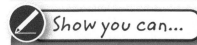 Show you can...

Complete this task to show that you understand atomic structure.

Draw out a timeline for the history of atomic theory. Include date, names and a summary diagram for each key development.

2.3 Elements

As soon as John Dalton had come up with his early ideas on atoms he started creating symbols to represent the different types of atoms already discovered. Today chemists all over the world use an agreed set of symbols for all known elements.

● **Figure 1** These are some of John Dalton's chemical symbols. What benefits could be gained by having standard symbols for the chemical elements?

An **element** is a pure substance that cannot be split up into simpler substances. Each element contains only one type of atom.

These elements are listed in the **Periodic Table**. Every element has different properties such as hardness or melting point. You cannot make elements out of other materials.

→ Chemical symbols

Every element has a chemical symbol. Scientists use these symbols as a shorthand version of an element's name. The table shows some examples of elements and their chemical symbols.

The symbol for many elements is derived from either the first letter (e.g. H for hydrogen) or the first two letters (e.g. He for helium) of their names. A few elements have symbols derived from their Latin names, for example, *ferrum* for 'iron', which leads to the symbol of Fe. The Latin names were often determined based on the appearance of the element, for example, the Latin word *aurum* means 'yellow', which gives rise to the symbol Au for the element gold.

When you use a chemical symbol, it is important to write it correctly. If the symbol is just one letter, that letter must be a capital. If the symbol is two letters, then the first must be a capital and the second must be lowercase.

I ✓   i ✗

Ag ✓   AG ✗   ag ✗

Element	Chemical symbol
copper	Cu
sulfur	S
mercury	Hg
carbon	C
iodine	I
bromine	Br
gold	Au
lead	Pb
fluorine	F
chlorine	Cl
silver	Ag
oxygen	O
hydrogen	H

→ The Periodic Table

The place to look for the symbols of all the known elements is the Periodic Table. If a substance is shown on there, you can be sure it is an element. The table groups elements in many useful ways, one of which is the arrangement dividing metal and non-metal elements.

Some elements do not exist as single atoms; they combine to form elemental **molecules**. Examples of these include the non-metals, hydrogen, nitrogen and oxygen, plus all the non-metals in the column with fluorine (F) at the top.

These elements are called **diatomic** ('di' means two, so this means 'made of two atoms') and we represent them using a small '2' after the symbol. For example oxygen becomes O_2.

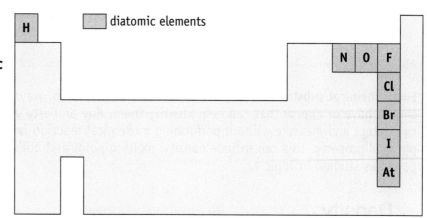

● **Figure 2** The Periodic Table divides the elements into metals and non-metals

● **Figure 3** The diatomic elements on the Periodic Table

? Questions

1 Name three elements on Dalton's original list (see Figure 1), which are no longer included on the modern Periodic Table.
2 Identify the following as metal or non-metal:
   a) iron (Fe)
   b) sodium (Na)
   c) chlorine (Cl)
   d) zinc (Zn)
   e) sulfur (S).
3 Why is it important to have a set of agreed symbols for all known elements?
4 Write out the symbols for the following elements:
   a) copper
   b) iron
   c) potassium
   d) fluorine
   e) silicon
   f) hydrogen.
5 Why do you think some of Dalton's symbols were changed?

✎ Show you can...

Complete this task to show that you understand about elements.

Copy and complete the table below. Fill in as many elements as you can. You may want to look up some melting and boiling points from other topics to help you.

State at room temperature		
Solid	Liquid	Gas

2.4 Measuring physical properties of elements

Since the start of modern chemistry, around 1650, chemists have been called upon to identify the substances that items are made from. By understanding the way different substances look and behave, chemists can carry out simple tests to identify them. They may also be able to work out the age of the item and even suggest some of the places it has been.

● **Figure 1** Why would it be useful to identify the metal this artefact is made from?

Every chemical substance has specific characteristics, that is, ways they behave or appear that can help identify them. Any property you can detect and measure without performing a chemical reaction is a physical property. This can include density, melting point and boiling point, as studied in Topic 1.

→ Density

Density measures how much mass there is in a fixed volume of a material. The higher the density of material, the more mass there is in each cubic centimetre (cm^3).

Density is measured in grams per cubic centimetre (g/cm^3). For example, if a material has a density of $2\,g/cm^3$, it means that each cm^3 has a mass of $2\,g$.

You can use a top pan balance to measure mass. Some balances are more accurate than others, because they have been made well and **calibrated** (compared with a standard mass).

- Choose the best quality instrument available.
- Set the balance at zero before you put the mass on.

● **Figure 2** Measuring mass using a top pan balance

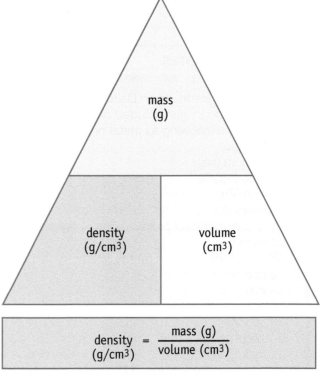

$$\text{density} \; (g/cm^3) = \frac{\text{mass (g)}}{\text{volume (cm}^3\text{)}}$$

● **Figure 3** The calculation for density

→ Finding the volume of a solid

To find the volume of a solid that has a cuboid shape, measure its length, width and height with a ruler.

> volume = length × width × height

You can use a measuring cylinder to measure the volume of any small solid with a regular or irregular shape.

1 Put a known amount of water into the measuring cylinder.
2 Gently place the object into the water. Make sure it is completely under water.
3 Measure the rise in volume.

> volume of object = (volume of water + object) − (volume of water alone)

For bigger objects, use a displacement can to measure volume.

1 Completely fill the displacement can with water. Add the object gently.
2 Catch the water displaced in a measuring cylinder.
3 Read its volume.

> volume of water displaced = volume of the object

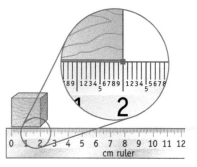

● **Figure 4** Measuring the length of a cuboid using a ruler

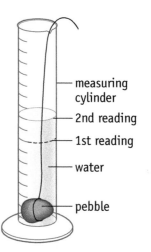

● **Figure 5** Measuring volume using the displacement of water method

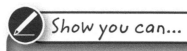

? Questions

1 Why is the string used in the apparatus in Figure 5?
2 Two objects each have a volume of $2\,cm^3$. The first has a mass of 6 g and the second a mass of 3 g. Which object has the greatest density?
3 A cube of iron has a length of 3 cm, a height of 1 cm and a width of 2.2 cm. What is its volume?
4 The equation triangle in Figure 3 helps you to work out the equation for any of the three properties shown. To do this, cover up the factor you are trying to work out and the triangle will show you the equation. Remember, the line across means divide and the line down means multiply. Complete the following equations using the triangle.
   a) density =
   b) mass =
   c) volume =
5 A student tested three different substances. Put the substances in order of increasing density.

Substance	Mass (g)	Volume (cm³)
A	32.3	1.7
B	1.25	0.5
C	38.4	3.2

✎ Show you can...

Complete this task to show that you understand how to measure physical properties.

Describe, with diagrams if that helps, how you could adapt the displacement method to measure volume for objects too big to fit in a measuring cylinder. Once you have chosen a container that is big enough, make sure you can collect any water that spills out.

Presenting and analysing data

→ Investigating density

Patrick and Holly have been learning about **density**. They have been shown some different methods for measuring density and have studied the ideas in Topic 2.4.

They have been asked to plan how to investigate the density of everyday objects. They collected some objects from home and their teacher gave them a couple of things from the Science lab. See if you have some similar objects in your pencil case so you can discuss and understand some of the challenges they faced.

They decided to try out both methods that they had learnt, so they could see if the data would compare.

Patrick wanted to start with choosing the objects that could be measured with a ruler in order to determine their volume.

1 Which of the objects would be suitable for this method?

Patrick measured the length, width and height of each object and found its mass using a top pan balance. His data was as follows:

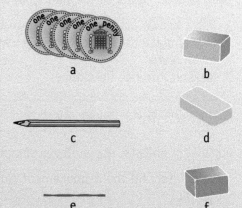

● **Figure 1** The collection of objects Patrick and Holly were investigating: a) five 1p coins, b) a cuboid of iron (1 × 1 × 3 cm), c) a pencil, d) an eraser (1.5 × 3 × 3 cm), e) a copper wire, f) a cuboid of copper (1 × 1 × 2.3 cm)

Object	Length (cm)	Width (cm)	Height (cm)	Volume (cm³)	Mass (g)
b	3	1	1	3	23.61
d	3	3	1.5	13.5	22.14
f	2.3	1	1	2.3	20.61

2 Patrick needs to add an additional column to his table for density. What unit would he need to include in this column? What is the equation needed to calculate density?
3 Copy and complete this table, adding in the additional column for density.

Holly decided to use the displacement method as shown on page 89. Patrick reminded Holly about the two hints their teacher had given them:

● Make sure the cylinder has an appropriate range and scale for your measurement. A cylinder with 1 cm³ scale divisions will give greater resolution than a cylinder with divisions only every 2 or 10 cm³.
● To get reliable results, be consistent – make sure you always measure to the bottom of the meniscus. Your eyes should be level with the meniscus.

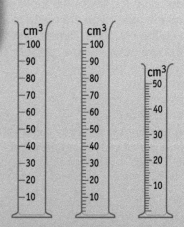

● **Figure 2** Different measuring cylinders offer different resolutions in measurement

For each object, Holly had to select the correct measuring cylinder; she had a choice of two measuring cylinders: a 25 cm³ and a 100 cm³ cylinder.

See if you can find one of each in your classroom to remind yourself of the size. Her selection would need to be based on which cylinder the object would fit into.

4 Which cylinder do you think she selected for each object?

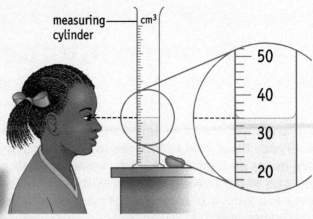

measuring cylinder — cm³

50
40
30
20

● **Figure 3** Crouch down to make sure you are reading the volume at eye level

Holly then used the displacement method. She recorded the start and end volume from the cylinder's scale. She also found the mass on a top pan balance.

Holly's data were as shown.

Object	First reading (cm³)	Second reading (cm³)	Mass (g)	Density (g/cm³)
a	15.0	17.5	17.80	
b	19.5	22.5	23.61	
c	50	56	5.44	
d	50	64	22.14	
e	15.0	16.5	11.00	
f	15.0	17.5	20.61	

5 Holly realised she needed an additional column for volume. What calculation would she need to do to complete this column?
6 Copy and complete the table, remembering to add the extra column for volume.

Once their data were collected they began thinking about explanations for some of their observations. Look carefully at the tables you have completed and compare the values for volume from Patrick's and Holly's data.

7 Why do you think Patrick's and Holly's values are different?

Objects e and f were both made of the same material: copper. Patrick and Holly had predicted that the density would be the same but when they compared the copper wire and the block of copper they found different values of density. They decided to look up the density of copper on the internet and found the value to be 8.96 g/cm³.

8 Describe the differences in density values calculated in Patrick's and Holly's experiments and the value they looked up.
9 Why do you think these were different?

3.1 Everyday acids and alkalis

We are surrounded by acids and alkalis in our everyday lives, and their uses in the chemistry laboratory are many and varied. By understanding their properties we can begin to identify them and start to explain why they are such a central part of our chemistry understanding.

● **Figure 1** The battery that starts the engine of this sports car is filled with acid. Which acids do you use in everyday life?

The word **acid** means 'sour tasting'. If you have ever enjoyed the citric acids found inside oranges and lemons you will know why. Our taste buds are designed to detect sour tastes.

An **alkali** is a substance that is the chemical opposite of an acid.

Other effects of acids and alkalis make them useful. For example, most bacteria, some of which can make our food go off, cannot grow in very acidic environments. Alkalis are able to break down grease and fats, allowing them to be washed away in water.

Even our own bodies have a range of acidic and alkaline substances. For example, your skin is slightly acidic; shampoos and shower gels that are 'pH balanced' have the same pH (level of acidity) as skin, which is important to keep your skin healthy. Our blood and stomach are acidic, whereas our tears are very slightly alkaline.

→ Acid or alkali?

You cannot tell just by looking at an item whether it is an acid, an alkali or a **neutral** substance (one that is neither acidic nor alkaline). One possible way to know is to look at labels, as they may show the inclusion of an acid or alkali. Canned and packet foods have these, as well as cleaning products.

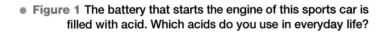

> Diet Cola
> Naturally flavored soda mix. No aspartame.
> No preservatives. Very low sodium. Contains caffeine.
>
> **Ingredients:**
> Water, Caramel, Coloring, Sucralose, Acesulfame
> Potassium, **Phosphoric Acid**, Natural Flavor,
> Sodium Citrate, Caffeine.
> Contains Sulfites.

● **Figure 2** The ingredients listed on foods in our cupboards will include a range of acids

Testing with litmus or Universal Indicator

You can also carry out a simple chemical test to find out whether a substance is an acid or an alkali. **Litmus** is a coloured dye, which will change colour in acidic or alkaline conditions; it is an example of a simple indicator. Litmus dye is soaked into some paper and dried; it can then be dipped into substances for testing. It comes in two types: blue litmus paper for testing for acidity and red litmus paper for testing for alkalinity. To test solid substances you need to dampen the paper first.

You could test a range of harmless items in the home using this method. Some possible results are shown in the table:

Substance	Result
cola	acid
bottled water	neutral
yoghurt	acid
lemon juice	acid
salt	neutral
baking soda	alkali
tomato sauce	acid
soap	alkali
toothpaste	alkali
pickled onion juice (vinegar)	acid
indigestion remedy	alkali

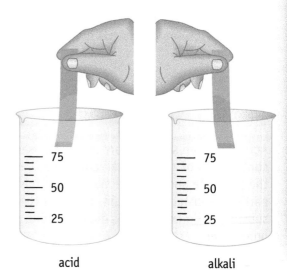

acid   alkali

● **Figure 3** Litmus paper can show if a substance is an acid or an alkali

Another example is **Universal indicator**; this gives a wider range of colours and can indicate the level of acidity. You can read more about this on page 94.

Some of the acids and alkalis in our homes are too dangerous to be tested without proper safety equipment, including eye protection. If they were brought into the lab you would collect the following results:

Substance	Result
oven cleaner	alkali
bleach	alkali
car battery fluid	acid
washing powder	alkali

? Questions

1 Name two acidic substances, two alkaline substances and two neutral substances.
2 What would be the difference in how you would test vinegar and soap with litmus paper?
3 Why do vegetables last longer if we pickle them in vinegar?
4 Would you expect lemonade to be acidic or alkaline and why?
5 What are the limitations of testing substances with litmus paper?

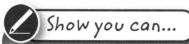

Show you can...

Complete this task to show that you understand the importance of acids and alkalis in our everyday lives.

Write a diary of your typical day. Include details of all the substances you use or see used. Go through your account and highlight or underline any acidic substances in red, alkaline substances in blue and neutral substances in green.

3.2 Indicators and pH

The amount or concentration of acid or alkali is an important factor to consider. Knowing this allows us to plan safely and track pH changes during any chemical reactions. In order to measure the acidity or alkalinity of a substance we use a pH indicator or pH meter.

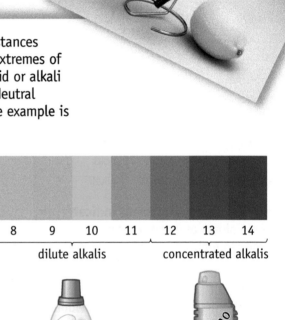

● **Figure 1** Why do you think a fruit battery will not last as long as a car battery?

Acid and alkali solutions are not all the same. Some substances contain lots of acid or alkali particles. These are at the extremes of the **pH scale** and they are more dangerous. The fewer acid or alkali particles substances contain, the safer they are to use. Neutral substances do not show acidic or alkaline properties; one example is pure water.

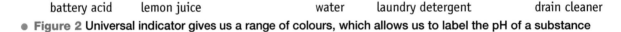

| pH | 0 | 1 | 2 | 3 | 4 | 5 | 6 | 7 | 8 | 9 | 10 | 11 | 12 | 13 | 14 |

concentrated acids   dilute acids   neutral   dilute alkalis   concentrated alkalis

battery acid   lemon juice   water   laundry detergent   drain cleaner

● **Figure 2** Universal indicator gives us a range of colours, which allows us to label the pH of a substance

→ Acids

- have a low pH – less than 7
- can be corrosive
- are neutralised by alkalis
- the most concentrated acids have the lowest pH

These are some of the acids you will use in the school lab:

- hydrochloric acid
- nitric acid
- sulfuric acid.

→ Alkalis

- have a high pH – more than 7
- can be corrosive
- are neutralised by acids
- the most concentrated alkalis have the highest pH

Alkalis you may come across in the lab include:

- sodium hydroxide
- ammonia
- sodium bicarbonate.

→ Using indicators

In the laboratory we use two main indicators, litmus and Universal indicator.

- Acids always turn litmus red.
- Alkalis always turn litmus blue.
- The limitation of litmus is that it cannot tell you how concentrated the acid or alkali is, so cannot be used to give a pH value.
- Universal indicator is a mixture of dyes that changes to different colours according to how concentrated the acid or alkali is. It gives the range of colours shown in Figure 2, which can be used to give a pH value.

→ Making indicators

Some plants contain substances that can be used as indicators, as they change colour in acids or alkalis. In order to make the indicator useful it must be extracted or removed from the plant cells.

1. red cabbage is cut up and ground to push out the juice

2. warm water is added

3. the juice and cabbage mixture is filtered to give an indicator that is ready to use

● Figure 3 Making an indicator

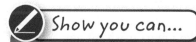

● **Figure 4 Why do you think we do not always use pH meters?**

You can also test for acidity or alkalinity with an electronic pH meter. The advantage to these meters is the resolution. As you can see in Figure 4 it gives a pH value to two decimal places. If you are trying to see a very small change in pH, a meter would be essential.

? Questions

1 What is an indicator?
2 Name a substance that would have a pH below 7, a pH of 7, and a pH above 7. Label them as acid, alkali or neutral.
3 What is the advantage of determining a pH value over simply recognising if a substance is acidic or alkaline?
4 Some people have lemon in their tea as opposed to milk. When the lemon is added, the tea changes colour slightly. Suggest why this may happen.
5 Explain each stage in the extraction of indicator from red cabbage.

Show you can...

Complete this task to show that you understand the pH scale and the range of acidic and alkaline substances.

Copy the Universal indicator pH scale as shown on page 94. Add as many examples as you can for each pH section. Use the ones here plus others from throughout this topic; you may study some in the lesson and you could research some further examples on the internet.

3.3 Dilution and safety

Acids range from the highly corrosive acid in car batteries to the flavoursome lemon juice added to salads and fruit, so understanding the hazards of acids and alkalis is essential both in the laboratory and in everyday life.

● **Figure 1** How can one acid be so dangerous it requires high-level safety equipment yet another is safe enough to eat?

→ Laboratory safety

Keeping safe in a chemistry laboratory is one of the first things you must learn. All of the safety rules apply to working with acids and alkalis as these could be some of the most dangerous chemicals you use.

When working with acids and alkalis, we need to understand their particular hazards and how to reduce the risk.

Dilution

- **Concentrated** acids and alkalis are **corrosive** – they can destroy a range of substances including metals and living tissue.
- **Dilute** acids and alkalis are not corrosive but they may be labelled with the warning sign for moderate hazard. This means that they will make your skin red or blistered and damage your eyes.
- You must wash off any acids and alkalis splashed on your skin, using plenty of water.
- You must wear eye protection.

● **Figure 2** Corrosive   ● **Figure 3** Moderate hazard   ● **Figure 4** Wear eye protection

- Acids and alkalis can be made less concentrated by diluting them with water. This is the reason that washing your skin with plenty of water will reduce the damage caused by an irritant.
- When we use acids and alkalis in reactions it is important we know how concentrated it is. We need to be precise when diluting.

Making a dilution

1 Add 90 cm³ of water to a clean beaker.
2 Add 10 cm³ of the starting solution.
3 Stir to ensure even mixing.

A concentrated acid can also be made safer by neutralisation. You can read about this on page 98.

Temperature

Another factor that affects the hazards associated with acids and alkalis is their temperature. If you have planned an investigation that requires heating an acid or alkali, the safety precautions must be greater. This is because the hotter the acid or alkali, the greater the corrosion risk. Even if the substance you have chosen is labelled an irritant, assume that by heating it you will need to take the same precautions as for a corrosive substance.

transfer 10 cm³ acid

add 90 cm³ water

100 cm³ acid   stir mixture   100 cm³ dilution

● **Figure 5** This shows a ten-fold dilution. This would result in a change of one pH unit

→ Transportation and storage

When acids and alkalis are transported or stored, safety precautions must be considered. In Europe, lorries transporting dangerous chemicals must display an orange warning sign. In case of an accident, the emergency services will need to know about the nature of the chemicals on board. You will recognise the familiar warning sign for a corrosive substance. The number codes give specific references, the top number indicating a more detailed hazard description and the bottom number representing the exact chemical inside the lorry. A key is needed to interpret these codes.

In school, the lab technicians will prepare the acids and alkalis to the right dilutions for use in class. In the preparation room, there will be very concentrated, and therefore corrosive, acids and alkalis and these must be stored safely in a dedicated cupboard, which is locked. Plastic trays may be used to contain any spills.

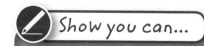

● **Figure 6** Why do you think codes are used to describe the hazards when transporting chemicals?

? Questions

1 Describe the difference between corrosive and irritant.
2 Why do you think eye protection is so important when working with acids and alkalis?
3 Why do you think it is important for the emergency services and doctors to understand the hazards associated with chemicals?
4 If 10 cm³ of concentrated acid with pH 1 was added to 90 cm³ of water, what would you expect the final pH to be?
5 If dilute acids and alkalis are safer, why do you think concentrated versions are transported and purchased in schools?

✎ Show you can...

Complete this task to show that you understand the importance of working safely with acids and alkalis.

Explain how each of the following laboratory safety rules can reduce the hazard of using an acid:

• eye protection
• tying hair back
• immediately washing chemicals off skin
• standing during all experiments.

3.4 Neutralisation

Another way to affect the pH of an acidic substance is by reacting it with an alkali. This chemical change is called neutralisation and its applications can range from indigestion remedies to neutralising acidic lakes.

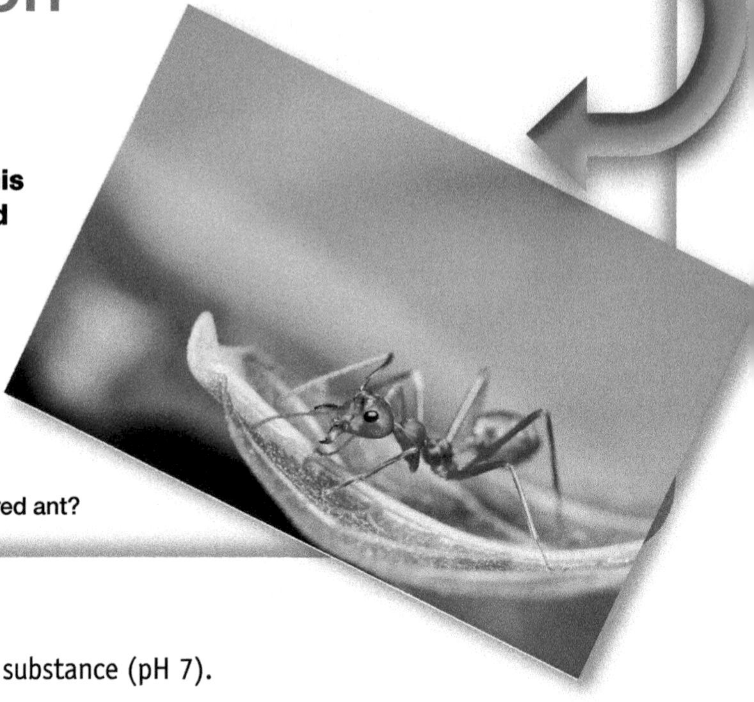

● **Figure 1** How could you soothe the acidic sting of a red ant?

→ Neutralisation

Acids and **alkalis** react together to form a **neutral** substance (pH 7). This reaction is called **neutralisation**.

The neutral substances formed are called salts.

acid + alkali → salt + water

Common salt (sodium chloride) is just one example of a salt. All salts have two-part names like this:

- sodium chloride
- silver bromide
- magnesium sulfate.

Examples of using neutralisation include the following.

- Indigestion remedies – these contain carbonates, which are alkalis. The alkali neutralises some of the stomach acid.
- Farmers and gardeners add alkalis to acidic soil to make the soil slightly alkaline and encourage plant growth.
- Neutralising lakes that have been affected by acid rain.

The acid and alkali have to be added in just the right amounts. If you add a dilute alkali to a concentrated acid you will only neutralise some of the acid. You would then have to add more alkali to get a neutral solution.

● **Figure 2** What evidence can you see here to show that a chemical reaction is taking place?

→ Investigating indigestion remedies

You could use the idea of neutralisation to compare how effective different indigestion remedies are. Indigestion is caused when some of the acid comes out of the stomach and into the gullet. There is no protection here and so the irritant acid causes pain.

- When all the 'stomach' acid has been neutralised the pH will be 7.
- An indicator can show when this happens.
- A pH meter will give continuous readings.

pH is a continuous scale so a pH meter can give more precise measurements. If the pH meter is attached to a datalogger you can take frequent measurements and track the changes in pH throughout the reaction.

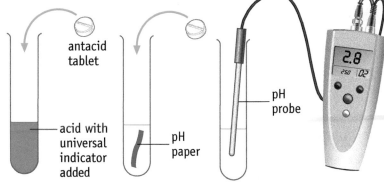

● **Figure 3** By adding indicator solution or paper, or using a pH meter, you will be able to see when the mixture has become neutral

Variables that could have an effect are:

- type of indigestion remedy
- amount of indigestion remedy
- form the indigestion remedy comes in – powder or tablets?
- pH of 'stomach' acid
- amount of 'stomach' acid.

You could test your ideas by measuring:

- how much indigestion remedy it takes to neutralise the 'stomach' acid
- how much 'stomach' acid is neutralised by a set amount of indigestion remedy
- the change in pH when a set amount of indigestion remedy is added to the 'stomach' acid.

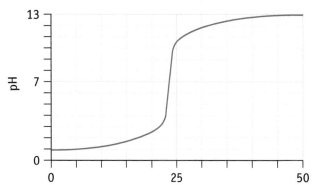

● **Figure 4** Using a pH meter attached to a datalogger can give a detailed plot of pH changes throughout a reaction

Remember that to plan an effective strategy you must make sure that you change only one variable, called the independent variable. All the other variables, known as control variables, need to be kept the same every time.

Remember to wear eye protection when working with acids.

Show you can...

Complete this task to show that you understand the process of neutralisation.

Think about an investigation into the effectiveness of an indigestion remedy. You will have to identify possible factors to investigate first and create a testable hypothesis.

For each of the following factors write a testable hypothesis:

- whether the remedy is a powder or a solid
- how much acid there is to neutralise
- the type or brand of indigestion remedy used.

? Questions

1. Describe what happens in a neutralisation reaction.
2. You have decided to use universal indicator. If you start with an acid and slowly add an alkali what colour changes would you expect to see?
3. What volume of indigestion remedy was needed to neutralise the acid in Figure 3? Do not forget to include the units.
4. Why would drinking plenty of water be a good idea for someone who is suffering from indigestion?
5. The pH meter is more precise. Explain what this means and outline the advantage of planning a strategy for an investigation that uses more precise instruments for measuring the dependent variable.

Making and recording observations

→ Soothing the ache of heartburn

Indigestion is often described by sufferers as a burning sensation around the heart. What is actually happening is that the acidic contents of the stomach have leaked out and are sitting in the gullet. As this area has no protection against acidic conditions, it causes a burning sensation. As this occurs near the heart, people mistake the pain.

A common remedy for heartburn is to take medicines designed to neutralise the acid and sooth the burning feeling.

The investigation

Two students, Steven and Julie, decided to plan a strategy for investigating the effectiveness of different brands of indigestion tablet. Their plans and some data they have collected are presented here.

The **hypothesis** the students created was this: the more expensive the brand of indigestion tablets the more hydrochloric acid they will neutralise.

Here is their method:

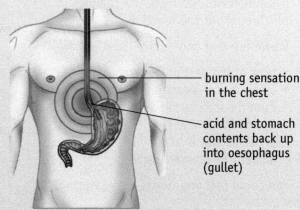

● **Figure 1** The uncomfortable feeling of burning during indigestion is caused by leaking stomach acid

1 Wearing eye protection, take a whole tablet from the first brand and grind it in the pestle and mortar.
2 Add this powder to a 100 cm³ beaker.
3 Measure out 10 cm³ of hydrochloric acid and add this to the beaker.
4 Using a piece of red litmus paper, test if the mixture is alkali.
5 Keep adding the acid in 10 cm³ portions until the litmus paper turns blue.
6 Record the total volume of acid added.
7 Repeat with a second and third brand of indigestion tablet.

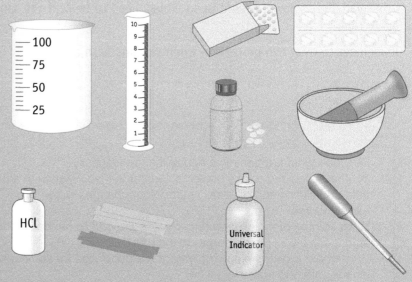

● **Figure 2** The selection of equipment available to Steven and Julie

Your task

1 Draw out a series of diagrams to show each step of Steven and Julie's investigation.
2 Suggest a change to step 4 and explain your reasons.
3 State the limitation of adding 10 cm³ portions each time.
4 Suggest an improvement to the method that would allow them to measure the total volume of acid neutralised.

Here are Steven and Julie's results:

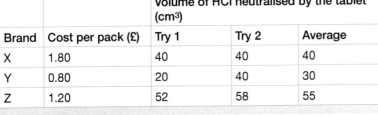

Brand	Cost per pack (£)	Volume of HCl neutralised by the tablet (cm³)		
		Try 1	Try 2	Average
X	1.80	40	40	40
Y	0.80	20	40	30
Z	1.20	52	58	55

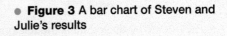

● **Figure 3** A bar chart of Steven and Julie's results

5 Why do you think they repeated each experiment twice?

When we collect data, we need to look for any repeat measurements that do not match. These are called outliers. In order to check the reliability of the experiment you would repeat the test and determine a third point. If data are still very different you continue to repeat until you have at least two points, that are close in value.

6 Identify the set of data most in need of further repeats. Explain your choice.
7 What are the limitations of presenting the data in a bar chart as in Figure 3?

Steven and Julie's teacher advised them to make some changes. She suggested they carry out a third repeat and reminded them that if there is an outlier it does not get included in the calculation of the average. She also asked them to plot a line graph of cost per tablet against volume of HCl neutralised.

Steven and Julie repeated the experiment a third time. Brand X gave 40 cm³, Brand Y gave 20 cm³ and Brand Z gave 60 cm³. They also needed to work out the cost per tablet. Brand X contained 36 tablets per pack, Brand Y contained 24 and Brand Z had 12.

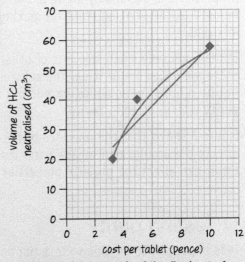

● **Figure 4** A line graph of the final set of data showing two possible lines of best fit

8 Draw out a results table showing all the data they have now and recalculate the averages.
9 Why did the teacher ask them to work out the cost per tablet?
10 Which line of best fit in Figure 4 do you think is best?

Pure and impure substances

The term 'purity' is often used in everyday language but chemists have a different definition. Most naturally occurring chemicals that might be called 'pure' in everyday language are in fact complex mixtures of many different types of atoms. Just as we can define an element by its inclusion in the Periodic Table, we recognise purity of a substance as it being made up of only one type of substance.

● **Figure 1** Why do you think the designers would choose pure gold to make this object?

Mixtures, or impure substances, contain two or more different substances that are not chemically joined together. These substances can be separated by physical means.

Pure substances contain only atoms or molecules of that substance. Their physical and chemical properties can be reliably predicted and they can be used as the starting point for making useful mixtures.

If there are only small amounts of other substances these are called **impurities**. The substance will not be referred to as a mixture but will be given a purity rating, often % purity.

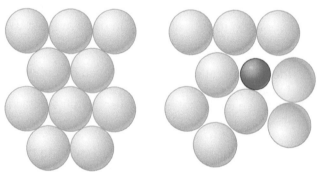

● **Figure 2** Pure gold, containing only gold atoms; and 90% pure gold as 9/10 atoms are gold

Figure 2 shows that the presence of impurities affects the arrangement of the particles and as a result can affect the physical properties of the substance.

→ Testing purity

It is important to be able to test the purity of a substance. This is done by measuring the physical properties of the substance. There are reliable data sources which quote the physical properties of pure substances. For example pure water will freeze at 0 °C and boil at 100 °C. If there are impurities in the water these values will change, giving an indication that the sample is not pure. Another method would be to determine the density of the substance and compare this to the data.

Impurities can be added in order to influence the physical properties of pure substances. For example, salt is used in grit for icy roads as it lowers the freezing point and melts the ice.

Chromatography

Another way to assess purity is to use **chromatography**. This is particularly useful to assess the purity of a coloured substance.

Chromatography works because the different dye particles dissolve and travel along the damp chromatography paper at different rates. If a substance is pure then a single dot will be present on the chromatogram.

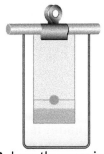

1. place a drop of the substance for testing on a piece of chromatography paper

2. hang the paper in a beaker of water. The water level should be below the drop of substance

3. as the water rises it takes the coloured substances with it. Multiple dots show a mixture

● **Figure 3 The key steps in paper chromatography**

→ Making useful mixtures

Chemists often combine pure samples in fixed combinations in order to create new and useful mixtures. A good example of this is the development of metal alloys. Particle diagrams can help to show us how mixing different metals together can alter the metal properties.

When chemists are designing materials, there is always a balance to be achieved. Generally increasing one property of the substance will lead to a reduction in another.

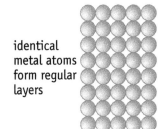

identical metal atoms form regular layers

the layers can slide easily over each other – metals are malleable and can be moulded

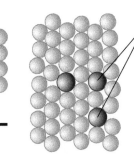

atoms of a different metal are added to create an alloy

the irregular pattern means the atoms can't slide easily – the alloy is harder and less malleable

● **Figure 4 The regular arrangement in pure metals explains why they are soft and malleable**

● **Figure 5 The bigger particles of the additional metal disrupt the regular pattern**

Questions

1 Why is chromatography most useful for analysing coloured substances?
2 Give an example of an object made of metal, which shows malleability.
3 If a substance is said to be 80% pure what would that mean in terms of particles?
4 Metals are often mixed to make them lighter by decreasing their density. Give two possible reasons why we may want the hardness of a metal but a lower mass.
5 Forensic scientists sometimes use chomatography to investigate paints left at a crime scene. What could this tell them?

Show you can...

Complete this task to show that you understand how the purity of metals affects their properties.

Metals are ductile. This means they can be drawn out into very thin wires. Based on Figure 4, create a diagram to describe how this could happen and explain why this would be easier with a pure sample of the metal.

4.2 Dissolving and solutions

Approximately 70% of our planet is covered in sea water, a complex mixture, largely made up of a salt mixed with water. The particles of each component are thoroughly combined with no visible indicator of the chemistry that has occurred when it formed.

● **Figure 1** Why does car paint not run off in the rain like the paint in this painting?

→ Dissolving

When a solid **dissolves** in a liquid, its particles slip into the gaps between the particles of the liquid. This means that the two different types of particles become completely mixed up.

For example, sugar will dissolve in water. When the sugar is first added you can see the crystal, but if it is left for a time or stirred the sugar particles will mix with the water creating a colourless solution.

- **Solute** – the solid that is being dissolved
- **Solvent** – the liquid the solute is being dissolved in
- **Solution** – the liquid that is formed when the solute has dissolved in the solvent

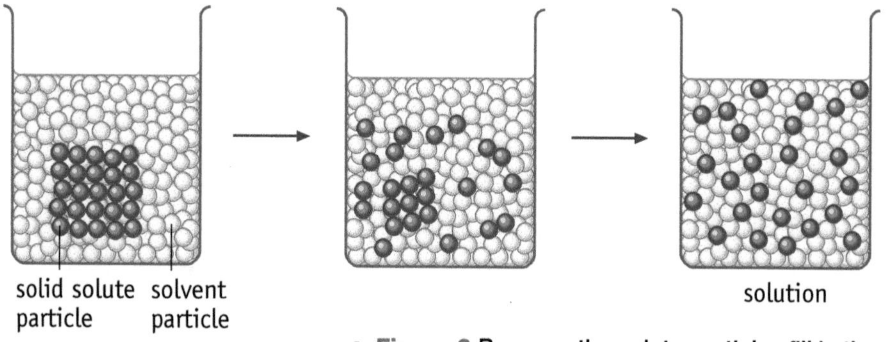

solid solute particle   solvent particle

solution

● **Figure 2** Because the solute particles fill in the gaps between the solvent particles, there is no increase in volume

Solids that dissolve are **soluble**. Solids that do not dissolve are **insoluble**. A solution is described as **saturated** when it contains as much solute as it can hold. If you keep adding solute to a saturated solution it will not dissolve.

→ Conservation of mass

Total mass does not change when a solid is dissolved in a solvent. The solution will have a mass that is equal to the mass of the solute added to the mass of the solvent. This is because you have the same number of particles at the end as you had to start with – they just get mixed up.

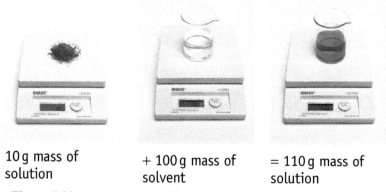

10 g mass of solution   + 100 g mass of solvent   = 110 g mass of solution

● **Figure 3** Mass is conserved (stays the same) when solute and solvent are combined

→ Dissolving can be reversed

Dissolving is a physical change so it can be reversed.

For example, if you leave salt (or sugar) solution to evaporate, the water particles gradually leave the solution and the salt (or sugar) particles will be left behind.

Water is not the only liquid we can dissolve things in. Many different liquids are used as solvents. Some solids are insoluble in water but soluble in certain other liquids.

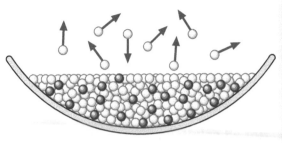

● **Figure 4** Like all physical changes, dissolving can be easily reversed

- Gloss paint will not dissolve in water so you need to clean the paint brush in a solvent like turpentine.
- Nail varnish will not dissolve in water so you need to use nail varnish remover. This usually contains a solvent called acetone.

→ Concentration

As you cannot tell just by looking how much solute has dissolved in a solvent, we need to be able to measure the amount of solute that goes into making up a solution.

Concentration tells us how much solute is dissolved and is measured in g/dm^3 (grams per decimetre cubed).

1 $1000\,cm^3$ of solvent will be measured in a cylinder; in this example the solvent is water. $1000\,cm^3$ is the same as $1\,dm^3$.
2 A set amount of solute will be measured on an electronic balance. Here this is shown as $10\,g$.
3 In a beaker large enough for the solvent, the solute and solvent will be combined.
4 The resulting solution will have a concentration of $10\,g/dm^3$ (10 grams per decimetre cubed).

● **Figure 5** How to make a solution of known concentration

Liquids can also mix together to make a solution. Two liquids that will mix together are called **miscible** and an example would be alcohol and water. Some liquids cannot mix, such as oil and water, and these are called **immiscible**. As liquids are measured in terms of volume and not mass, the concentration is then described in terms of cm^3/dm^3.

Show you can...

Complete this task to show that you understand what is happening to the particles in a solution.

In a salt solution, you cannot tell if there is any solute present. Draw a particle diagram of a salt solution and pure water.

Show that you know how to test a solution by planning a strategy you could use to show the liquid you have is salt solution and not pure water.

? Questions

1 If you mix sugar and water, what is the name of the:
   a) solute   b) solvent   c) solution?
2 What would you see if some sea water was left to evaporate?
3 If you dissolved 6 g of salt in $500\,cm^3$ of water, what would the concentration be in g/dm^3?
4 If 5 g of solute were dissolved in $100\,cm^3$ of water what would the end volume of the solution be? Explain your answer.
5 You have been given a salt solution of unknown concentration. How could you determine the concentration of the solution?

4.3 Factors affecting solubility

Not all solutions are the same! Beyond defining a substance as soluble or insoluble we also see a difference in how much of each substance can dissolve. This is defined as solubility, which can be affected by the nature of the solvent, the temperature and the surrounding pressure.

● **Figure 1** Why is it important that gases can dissolve in water?

Solubility is a physical property. It is a measure of how soluble a substance is, because it measures how much of a solute (or a gas) can dissolve in a fixed volume of **solvent** to make a saturated **solution**.

→ Solubility of solids in water

Water is an excellent solvent that dissolves many different things. Some substances dissolve in water better than others. You will notice that some substances do not dissolve at all, such as sand, whereas others have varying levels of solubility.

Another factor that affects solubility in water is temperature. Generally the higher the temperature of the water, the more solute will dissolve.

We can plot this variation and create a solubility curve; you can then use this to predict how much solute will dissolve at a set temperature.

This explains why as saturated liquids cool the solute comes out of solution and forms solid particles. For example, 100 cm³ of a saturated solution of sodium chloride at 80 °C contains 38 g of salt. As it cools to 30 °C only 36 g can dissolve. The difference of 2 g will come out of solution and will be seen as solid salt crystals at the bottom of the solution.

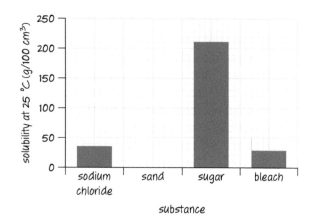

● **Figure 2** Bar graph to show varying solubilities of everyday substances in water

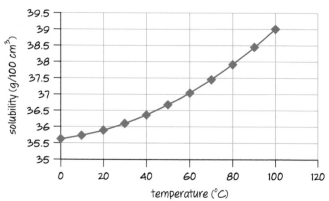

● **Figure 3** Solubility of the salt sodium chloride with varying temperature

106

→ Solubility of gases in water

The solubility of gases in water is vital for all living things. As humans, we transport much of our waste carbon dioxide back to the lungs dissolved in our blood; animals that live in water rely on the solubility of oxygen for their survival.

Gases dissolve in water to varying degrees but interestingly follow an opposing trend to that seen for solids. The higher the temperature of the water, the less gas will dissolve. This is why fizzy drinks go flat more quickly on a warm day as the gas slowly leaves the solution.

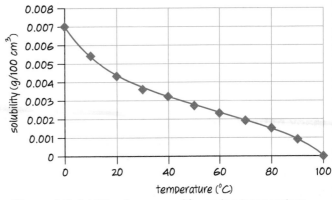

● **Figure 4** Solubility of oxygen with varying temperature

→ Investigating solubility

You could plan an investigation into the effect of temperature on solubility. Remember that to carry out a valid comparison you must make sure that you change only one **variable**. All the other variables need to be kept the same every time. The factor that you change – the temperature – is the independent variable. The factor that you measure is the dependent variable, which would be mass. The controlled variables are the type and volume of solvent and the type of solute.

To make sure your results are reliable you should repeat your readings. You can then check that you are getting similar results.

To reach a conclusion, you could plot a graph of the dependent variable (*y*-axis) against the independent variable (*x*-axis). This would compare to the solubility curve in Figure 3.

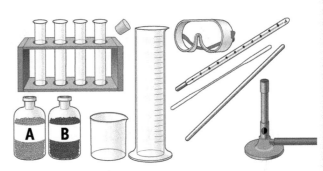

Temperature (°C)	Mass dissolved in saturated solution (g)	
	Try 1	Try 2
30	212	208
40	243	236
50	255	295
60	290	286
70	332	337

● **Figure 5** You should record all of your results in a table including the units in the top row of the table

Questions

1 Describe what you would see if another spoonful of solute was added to a saturated solution.
2 Why is it important that gases dissolve in water?
3 Look at the data in Figure 5. Which temperature should be repeated again to increase reliability?
4 If you heat 100 cm³ of a saturated solution of sodium chloride from 20 °C to 60 °C how much more salt will you be able to dissolve?
5 Explain what happens to a saturated solution when some of the solvent evaporates.

Show you can...

Complete this task to show that you understand the factors affecting solubility.

Explain each of the following observations:

• more salt dissolves in 100 cm³ of water than in 10 cm³ of water
• copper oxide is insoluble at room temperature but is soluble at 80 °C.

107

Many natural and man-made mixtures are found around us every day. In order to investigate and understand the nature of each part of a mixture, they must first be separated. The most accurate information will be found from the pure substance.

● **Figure 1** Why is it useful to be able to separate out the substances that make up a mixture?

→ Separation methods

Separation methods include:

- **filtering** – to separate insoluble solids from liquids
- **chromatography** – to separate a mixture of different compounds (e.g. dyes)
- **evaporation** – to separate a solute from a solvent
- **distillation** – to separate and collect the solvent
- **magnetism**
- **freezing**
- **melting**
- **fractional distillation**.

If only one of the substances in a mixture is magnetic, then it will be attracted towards a magnet but the rest of the mixture will not.

→ Freezing

A mixture of two or more different liquids can be separated by freezing. The different liquids freeze at different temperatures. The liquid with the highest melting point freezes first. The liquid with the lowest melting point freezes last.

In the example shown in Figure 3, the water freezes before the ethanol (alcohol).

● **Figure 2** Using a magnet to separate metal filings and sand

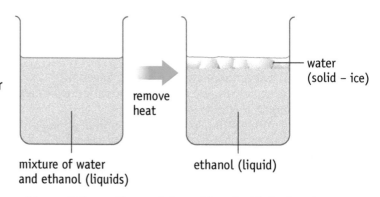

● **Figure 3** Separating a mixture of two liquids by freezing

water (solid – ice)

remove heat

mixture of water and ethanol (liquids)

ethanol (liquid)

→ Melting

A mixture of two or more different solids can be separated by melting. The solid with the lowest melting point melts first. Alloys are mixtures of two or more different metals. Alloys are melted to separate out the different metals. As each metal melts it is poured off into a separate container.

The freezing and melting methods work because every material has a set melting point and different materials have different melting points.

● Figure 4 Why do these workers have to wear safety goggles and gloves?

→ Filtration

Filtration, or filtering, is used to separate insoluble solids from liquids. Sand can be separated from a mixture of sand and salt water by filtering.

Filtering works because the small liquid particles pass through the filter paper into the container below but the insoluble solid cannot get through. The insoluble solid stays on the filter paper. For example, salt solution passes through the filter paper but sand does not.

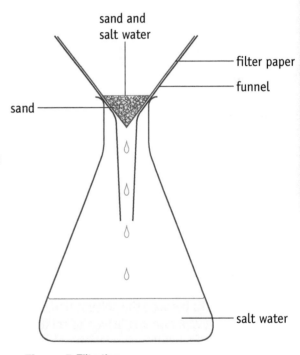

● Figure 5 Filtration

? Questions

1 What is the difference between soluble and miscible?
2 List all the state changes mentioned on these pages and describe what happens for each of them.
3 Explain why an insoluble substance will get blocked by the filter paper but a solute dissolved in water will pass through.
4 What effect would heating the mixture have on filtration?

✎ Show you can...

Show that you understand these separating techniques.

Draw a picture to show how melting a mixture of two solids could allow you to separate them. Use Figure 3 as a guide.

When a mixture includes a liquid, its components can be separated out by heating. A mixture of several different liquids can be separated by controlling the temperature so that only one substance boils at a time.

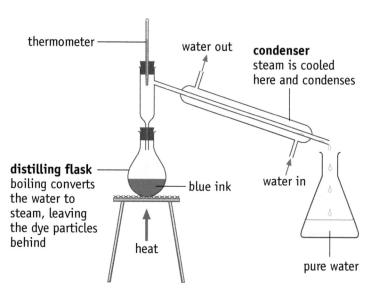

● Figure 1 How does the Sun help to produce pure salt?

→ Evaporation and distillation

Evaporation is used to separate a soluble solid from the liquid it is dissolved in. For example, evaporation can be used to separate salt from water in a salt water solution.

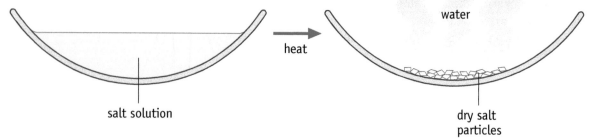

heat

water

salt solution

dry salt particles

● Figure 2 Evaporation

Evaporation works because the solvent particles in the solution evaporate and leave the solid (solute) behind. In simple evaporation only the solid can be collected because the liquid has evaporated away into the air.

Distillation is used when you want to separate and collect the liquid solvent. For example, pure water can be distilled from a salt water solution.

The solution is heated until it boils and the solvent particles turn into a gas. The gas is then cooled by the cold water running around the condenser. Cooling makes the gas condense back into a liquid. The liquid (pure solvent) is then collected. The solid is left behind in the original container so it can be collected too.

thermometer

water out

condenser
steam is cooled here and condenses

distilling flask
boiling converts the water to steam, leaving the dye particles behind

blue ink

water in

heat

pure water

● Figure 3 Distillation of ink and water

Fractional distillation separates out the different substances in a mixture of two or more liquids. It works because the different liquid 'fractions' have different boiling points.

1 The mixture is heated.
2 The fraction with the lowest boiling point boils first.
3 As it boils, it becomes a gas and travels up the tube and into the side arm.
4 It cools and condenses into a liquid.
5 The liquid is collected.

As the temperature rises, the fraction with the next lowest boiling point boils, condenses, and is collected. As each fraction boils it is collected in a different container.

Crude oil is a liquid. It is a mixture of many different compounds. These compounds are separated using fractional distillation.

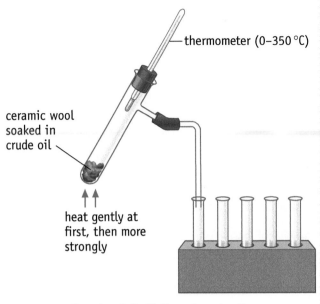

ceramic wool soaked in crude oil

thermometer (0–350 °C)

heat gently at first, then more strongly

● **Figure 4 Fractional distillation of crude oil**

→ Using the Bunsen burner

In the laboratory, we use the Bunsen burner to provide the energy needed for changes of state and many chemical changes. The temperature of a Bunsen flame can be as high as 2000 °C. You need to know the following safety points for using a Bunsen burner.

● Use the yellow flame (air hole closed) when the Bunsen burner is not being used.
● Ensure hair is tied back and no loose clothing is near the Bunsen burner.
● Keep the Bunsen burner well away from the edges of the desk, from people and from anything flammable.
● Use a heatproof mat.
● Wear eye protection.
● Always watch the Bunsen burner.

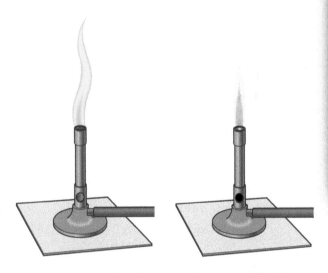

● **Figure 5 The yellow flame is easier to see. The blue flame is much hotter**

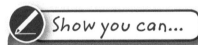 Show you can...

Complete this task to show that you understand how to select the correct equipment for experiments.

Draw the apparatus you would use to separate the following mixtures so you can collect a sample of both parts of the mixture:

• a mixture of blue, black and red ink
• the insoluble salt, silver chloride, and water
• salt solution
• sand and water
• a mixture of the two miscible liquids, ethanol and water.

? Questions

1 Give two reasons why the yellow flame of a Bunsen is called the safety flame.
2 What is the advantage of distillation over evaporation?
3 Draw a series of three particle diagrams, including a key, to explain what happens during distillation of ink and water.

Planning and designing investigations

→ Making a pure salt

Rachel and Louise were set the task to make a pure sample of the **salt**, copper sulfate. The first thing they did was to research some information about it.

Copper sulfate

Copper sulfate is a soluble salt; the salt and water combine to make a blue-coloured **solution**. It can be made by reacting copper oxide solution and sulfuric acid solution.

> 1 What is the **solute** and what is the **solvent** in the copper sulfate solution?

Copper oxide is an insoluble salt at room temperature but will dissolve at 80 °C. For the reaction to happen, a solution of the copper oxide was needed so they decided to add it to the acid and heat the mixture to get it to react.

Safety hazards

Louise remembered something about the hazards of acids so before they started they looked up some safety information. On the bottle of acid provided they saw the safety sign in Figure 1.

> 2 What does this sign stand for?

Based on this safety warning they added some steps into their plan, including wearing eye protection and working near a sink so they could wash any off if it spilt on their skin.

> 3 Explain why each of these precautions is necessary.
> 4 They are going to be heating the acid. Why does this make it even more important to be aware of the hazards?
> 5 Why have they decided to heat the acid?

Their teacher also advises them to make sure they stand up during the experiment.

> 6 Why would their teacher advise this?

Now the safety precautions are planned, they begin to gather their equipment.

> 7 Draw a diagram to show how they would set up their equipment to measure 20 cm³ of acid into a beaker and to heat this over a Bunsen burner. Add as many labels as you need.

● **Figure 1** The safety label on the sulfuric acid bottle

Rachel says they should stir the mixture as they add the copper oxide. Louise disagrees on safety grounds.

8 Why would Rachel's suggestion be helpful?

They agree to not stir, and begin to add the copper oxide. They add it until no more will dissolve because they want a saturated solution. They observe that a blue solution begins to form, starting as a very pale blue and getting darker as more copper oxide is added.

9 Why does the colour get darker?

Separation

Once no more copper oxide will dissolve they turn off the Bunsen burner and leave the solution to cool. They begin to see black powder collect at the bottom of the beaker and ask their teacher for help. They are told that this is just the extra copper oxide that did not react coming out of the solution, and they will need to plan a way to separate it from the copper sulfate solution.

● **Figure 2** What hazards can you see in this picture?

10 Why does the copper oxide appear as the solution cools?
11 How could they separate the copper oxide from the copper sulfate solution?

Once they have their copper sulfate solution they start to think about how to separate the salt from the water. Louise suggests distillation but Rachel wants to use evaporation.

12 Which student would you agree with and why?
13 Draw a diagram of the apparatus you would need for your chosen method.

The students now have a sample of pure copper sulfate.

14 How could they show their sample is pure?

● **Figure 3** A sample of pure copper sulfate

We know that compounds are formed from chemical reactions. Recognising and measuring chemical change is important, whether it be observing what happens when fuels burn in air, or the dramatic explosions of November fireworks.

● **Figure 1** How can you tell that a chemical change is taking place?

→ **Chemical and physical changes**

Chemical reactions make a new substance. This is called a chemical change. The change cannot be reversed, except in some cases by another chemical reaction.

Melting and **boiling** are physical changes. For example, changing ice to water or water to steam can be reversed – and no new substance is formed.

Here are some everyday changes that are chemical reactions.

- Iron reacts with oxygen in the air or in water to form iron oxide (rust) (Figure 2a). Other metals have similar reactions with oxygen, but may react faster or slower.

- When you bake a cake using flour, baking powder, eggs, sugar and margarine, new substances are formed. You cannot get the original ingredients back again.

- After burning, a match is black. Soot is one product of the burning reaction (Figure 2b).

- The chemicals in fireworks explode and produce coloured light.

- When you add baking powder or bath salts to water you see **effervescence** (bubbles) (Figure 2c).

Many useful new products are made using chemical reactions. Most plastics and many modern clothing fibres have been made from reactions using oil.

● **Figure 2** Some examples of chemical reactions

→ Reactants and products

The chemicals that react together are called **reactants**. The substance or substances formed are called **products**. Chemists write word equations to show what happens in a chemical reaction.

iron + oxygen → iron oxide
reactants   *product*

When a chemical reaction occurs there are many possible observations that indicate the rearranging of atoms into new substances.

- Colour change – the reactants look different from the products
- A change in temperature, or light is emitted
- Effervescence as bubbles of gas are produced, such as when acid is poured on a metal

● **Figure 3** Some signs that a chemical reaction has occurred

→ Measuring chemical changes

In many cases, it is not just useful to observe these changes but to also measure them in some way.

- Gases that are produced can be collected and their volume measured.
- Temperature changes during a reaction can be monitored using a thermometer.

Some reactions, such as those between acids and alkalis, have no observable changes. In these instances a pH indicator can be added to make the change in pH observable.

In some reactions, one of the products will be an insoluble solid. In these situations the reaction mixture will appear to go cloudy as the solid comes out of the solution. This is called a precipitation reaction.

put a cross underneath the beaker   the time taken for the cross to disappear can be measured

● **Figure 4** Observing a precipitation reaction

? Questions

1. Put the example reactions from Figure 2 in order from fastest reaction to slowest.
2. How could you measure the change in temperature as a reaction happens?
3. What signs of a chemical reaction can you list when fireworks explode?
4. The salt sodium chloride is made when sodium reacts with chlorine, and energy is released as light showing that this is a chemical change. Name the product and the reactants.
5. A student puts a piece of magnesium into some copper sulfate. The blue colour of the copper sulfate fades as magnesium sulfate is formed. A reddish-brown solid forms on the magnesium, indicating that copper is being made. Write a word equation showing the changes in this reaction.

✎ Show you can...

Show that you have understood chemical changes.

Describe all the observations you could make of a chemical change as you fry an egg.

5.2 Atoms and molecules in reactions

To understand chemical change, we must review the atomic nature of matter and consider the way atoms are combined in the starting substances and the new atomic arrangements in the products.

● **Figure 1** What happens to the atoms in sodium and chlorine as they react to make sodium chloride?

→ Rearranging atoms

When chemical reactions take place, the atoms in the **reactants** become arranged differently to form the **products**.

When two **elements** react together to form a **compound**, atoms of the elements combine. For example when calcium reacts with chlorine, the calcium and chlorine atoms bond together to make a compound called calcium chloride.

● **Figure 2** The rearrangement of atoms when calcium reacts with chlorine

The number of atoms in the reactants and the products are the same and therefore mass has stayed the same.

When two compounds react together, new bonds are formed between the atoms of the different elements that make up the compounds to produce the compounds that make up the product.

For example, a reaction takes place when hydrochloric acid is mixed with sodium hydroxide:

• the hydrochloric acid splits up into atoms
• the sodium hydroxide splits up into atoms
• the atoms in both compounds become arranged differently and form new bonds.

Sodium chloride and water are formed as products:

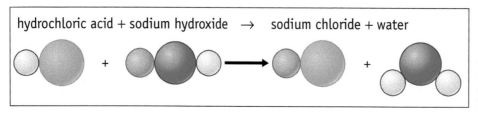

● **Figure 3** The rearrangement of atoms when hydrochloric acid reacts with sodium hydroxide

→ Molecules

Some elements exist as atoms and others as molecules. A few non-metals exist as molecules, made of two atoms, when they are in their elemental state. These diatomic molecules include hydrogen (H_2), nitrogen (N_2), oxygen (O_2) and the halogens (F_2, Cl_2, Br_2, I_2). They are molecular elements because both atoms are the same.

When a substance burns, it reacts with oxygen. When carbon is burnt, the carbon and oxygen atoms bond together to make a carbon dioxide molecule.

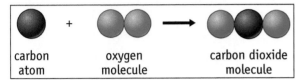

carbon   oxygen   carbon dioxide
atom   molecule   molecule

● **Figure 4** The rearrangement of atoms when carbon burns in oxygen

The carbon dioxide molecule has the same mass as the carbon and oxygen atoms added together. No atoms have been gained or lost.

When copper is burnt, the copper and oxygen atoms bond together to make the compound copper oxide:

copper + oxygen → copper oxide

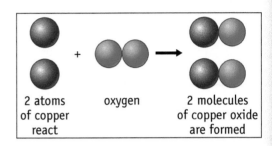

2 atoms   oxygen   2 molecules
of copper   of copper oxide
react   are formed

● **Figure 5** The number of atoms in the reactants is equal to the number of atoms in the products

→ Burning fuels

Fuels contain carbon and hydrogen.

- When fuels burn completely in oxygen, carbon dioxide and water are formed.
- Carbon and oxygen atoms bond together to make carbon dioxide molecules.
- Hydrogen and oxygen atoms bond together to make water molecules.
- The atoms are just rearranged to form new substances.

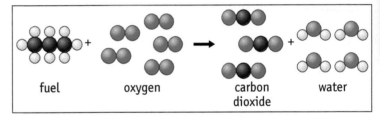

fuel   oxygen   carbon   water
                    dioxide

● **Figure 6** The rearrangement of atoms when a fuel burns in oxygen

When fuels burn without enough oxygen to react it is called **incomplete combustion** – the nature of the products changes. Instead of carbon dioxide (CO_2) forming, carbon monoxide (CO) forms. Water is also formed.

? Questions

1. What happens to the bond in hydrochloric acid when it reacts with sodium hydroxide?
2. What would you see if the rearrangement of atoms led to a gas being produced?
3. Describe what happens to the atoms as sodium reacts with water to form sodium hydroxide and hydrogen gas.
4. Which bonds need to be broken for hydrogen to react with oxygen? Which bonds are made?
5. Why are five molecules of oxygen required in the reaction shown in Figure 6?

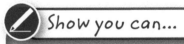

Show you can...

Complete this task to show that you understand the rearrangement of atoms during a chemical reaction.

Draw out a diagram, such as that in Figure 6, for the reaction between methane (the gas that is used to fuel Bunsen burners) and oxygen. Methane has the chemical formula CH_4 and its molecule has the carbon atom in the centre with the four hydrogen atoms surrounding it.

5.3 Testing gases

The production of a gas can be one of the easiest signs to help identify a chemical reaction. Often chemists want to identify the substances taking part in reactions, and identification of the gases produced is an ideal starting point.

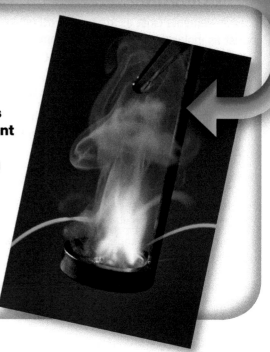

● Figure 1 How could you identify the gas being released here when potassium reacts with water?

→ Collecting gases

In order to identify the gas produced in a reaction it must first be collected. This can be achieved by blocking the tube with a bung. If the test to identify the gas requires it to be dissolved or mixed with a solution, a delivery tube can be used as in Figure 2.

At times chemists will want to know what gas they have and also how much of it has been made. In these situations more complex collection methods are needed where the amount of gas can be measured. Many gases can be collected over water as long as they are **insoluble**.

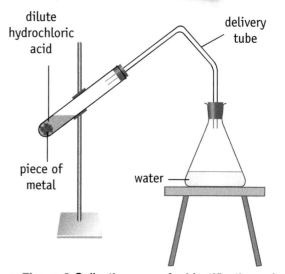

● Figure 2 Collecting gases for identification only

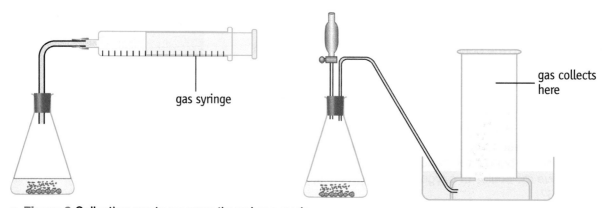

● Figure 3 Collecting gas to measure the volume made

→ Testing gases

Once the gas is collected, chemists can set about testing the gas that has been made. Usually they have an idea about which gas will be produced and so can choose the tests that are most appropriate. They can choose from bubbling the gas through another solution as in Figure 2 or testing with damp litmus paper or using a lit or glowing splint.

There are several gases that can be tested for easily in the laboratory; they include hydrogen, oxygen, chlorine, carbon dioxide, sulfur dioxide, ammonia and water vapour.

Each one has its own test and expected result.

some gases change blue litmus to red

a glowing splint relights in oxygen

● **Figure 4** Techniques for gas tests

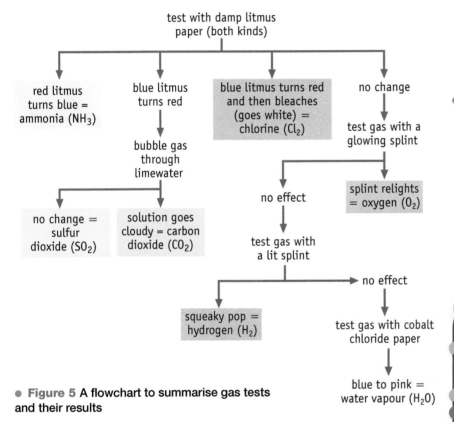

● **Figure 5** A flowchart to summarise gas tests and their results

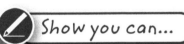

Show you can...

Complete this task to show that you understand how to collect and test gases.

Identify the following three gases based on the observations in the table.

Gas	Litmus	Limewater	Glowing splint	Lit splint	Cobalt chloride paper
X	no change	no change	went out	went out	blue to pink
Y	blue to red	no change	went out	went out	no change
Z	no change	no change	went out	squeaky pop	no change

Questions

1. What is the difference in testing for oxygen and hydrogen?
2. Which gases are neutral?
3. How could you identify a gas you have collected in a gas syringe?
4. Why would ammonia and carbon dioxide give the colour changes described in Figure 5?
5. Describe which gas collection method you would choose for the following situations:
   a) quick identification
   b) to know the volume and then to identify a gas
   c) to find the volume of a gas you know to be soluble in water.

5.4 Writing chemical equations

Not only do chemists write symbols for elements and compounds but they can use them to build up chemical equations. These show what happens when atoms rearrange during a chemical reaction. Chemical equations are a universal code that all chemists, wherever they work in the world can understand.

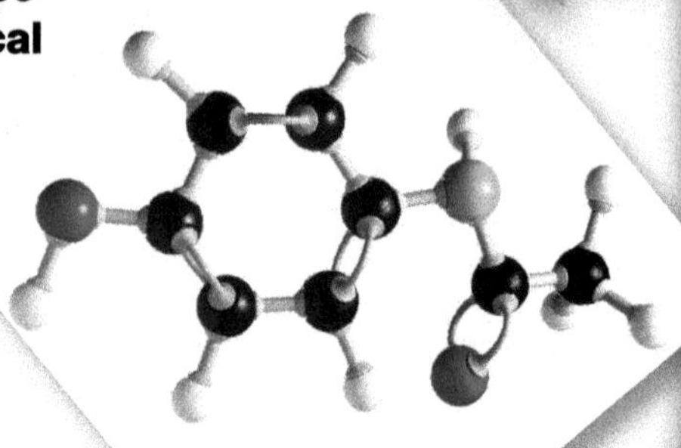

● **Figure 1** How would you represent this molecule using chemical symbols? (Hint: the blue atom is nitrogen.)

→ Word and symbol equations

Reactions can be represented as word equations or as equations with chemical symbols. Both types of equations are useful because they let you see what is happening in a reaction. Word equations are simpler to write than symbol equations, but scientists prefer equations using symbols.

Here is an example of a word and chemical symbol equation for the reaction between zinc and copper sulfate solution:

zinc + copper sulfate → zinc sulfate + copper
 Zn + $CuSO_4$ → $ZnSO_4$ +   Cu

When you use chemical symbols, the first letter of each symbol must be a capital – any others must be lower case (C, Ca, H, He).

Any numbers – like the '2' in CO_2 – must be written slightly below the letters. They are not written above, like in a 'squared' sign. The number applies only to the atom immediately before it.

CO_2 ✓   CO² ✗

→ Balancing equations

Balancing an equation means making sure you have the same number of each type of atom before and after the reaction has taken place. To make equations balance, numbers may have to be added in front of one or more of the symbols for the reactants and products. The number applies to all the atoms in the compound that follows.

The following equation for a reaction is not balanced:

$Mg + O_2 \rightarrow MgO$

There are two oxygen atoms on the left-hand side, but only one oxygen atom on the right-hand side.

The equation is balanced by adding a 2 in front of the MgO to balance the oxygen atoms, and a 2 in front of the Mg to balance the Mg atoms as there are now two atoms of Mg in the product:

$2Mg + O_2 \rightarrow 2MgO$

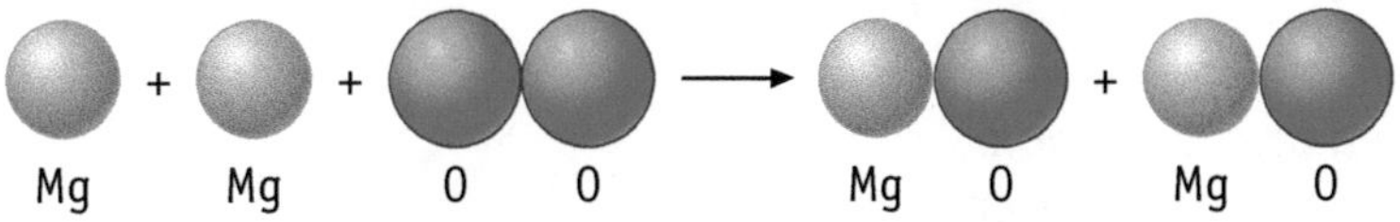

Mg   Mg   O   O   Mg   O   Mg   O

● **Figure 2** A particle diagram for the balanced equation above

A good way to get better at balancing equations is to move away from needing the particle diagrams to help you and simply add up the total number of atoms on each side of the equation. Each time you spot a number that is not equal you will need to add a number in front of the correct substance and then re-check the totals.

For example when calcium reacts with hydrochloric acid to make calcium chloride and hydrogen:

$Ca + HCl$ $\rightarrow$	$CaCl_2 + H_2$	
Ca = 1	Ca = 1	✓
H = 1	H = 2	✗
Cl = 1	Cl = 2	✗

The numbers of hydrogen and chlorine atoms are not balanced and so more hydrochloric acid is needed:

$Ca + 2HCl$ $\rightarrow$	$CaCl_2 + H_2$	
Ca = 1	Ca = 1	✓
H = + 2	H = 2	✓
Cl = + 2	Cl = 2	✓

? Questions

1. What is the difference between CO_2 and $2CO$?
2. How many oxygen atoms are there in the following compounds?
   a) Na_2SO_4 **b)** $Ca(OH)_2$ **c)** $Mg(NO_3)_2$
3. Balance this equation:
   $H_2 + O_2 \rightarrow H_2O$
4. Show, using a summary table like that for calcium and hydrochloric acid, that this equation is balanced:
   $Ca(OH)_2 + 2HNO_3 \rightarrow Ca(NO_3)_2 + 2H_2O$

✏ Show you can...

Complete this task to show that you understand how to write chemical equations.

Write a word and then a balanced symbol equation for the following reaction.

David took some magnesium oxide, an insoluble solid at room temperature. He heated it with some hydrochloric acid solution until is started to react. The resulting solution, in the water that was produced, contained the soluble salt magnesium chloride ($MgCl_2$).

Building scientific awareness

→ Investigating temperature change

Elsa and Jonathon wanted to investigate the temperature change during a chemical reaction. They had carried out several chemical reactions in class and noticed, by touch, that several felt warm during the reaction. They had noticed this in a previous topic when investigating the reaction of **acids**. They wanted to test whether the type of acid affects the temperature increase during a chemical reaction.

They wrote a rough plan before showing it to their teacher:

- take 25 cm³ of 1 M hydrochloric acid
- add a 3 cm length of magnesium ribbon and measure the temperature change
- repeat with 25 cm³ of 1 M sulfuric acid.

Before they began their teacher gave them one hint. **Repeatability**! Of course, they need to carry out each experiment at least twice to make sure the values are close. If not they can repeat again to check for any mistakes. Their teacher also gave a useful explanation about why repeats were needed, saying that it is to show **accuracy** and **precision**.

As Elsa was convinced that accuracy was all about the resolution of the thermometer, Jonathon asked her to research the options available and select the best one. She asked the lab technician what was available. Resolution is the smallest unit that can be distinguished on the equipment scale. For example, the number of divisions on a ruler or the number of decimal places on a digital scale.

accurate and precise

precise, but not accurate

not accurate, not precise

● **Figure 2** Bullseye – how accuracy and precision relate to hitting the true value

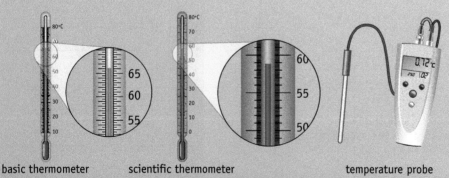

basic thermometer   scientific thermometer   temperature probe

● **Figure 3** The three options for measuring temperature

Elsa wanted to summarise the resolution of the three choices and so created this table:

Thermometer	Range (°C)	Scale division (°C)	Error (°C)
basic	−10 to 105	0.5	± 0.25
scientific	−5 to 80	0.1	± 0.05
electronic	unlimited	0.01	± 0.005

Jonathon did not understand the final column. Error of an instrument is a measure of uncertainty about the value read. Every piece of measuring equipment has it and the symbol ± can be read as 'plus or minus'. Elsa explained to Jonathon: 'if you have read a value on a basic thermometer, say 26 °C, it could actually be 25.75 (26 minus 0.25) or 26.25 °C (26 plus 0.25), or anywhere in between'.

1 What could the temperature be between if you read 26 °C on a scientific thermometer?

So for the best resolution they decided to use the electronic temperature probe. Jonathon then insisted they consider the importance of experimental technique on accuracy and precision and wanted to add more detail to their plan. He felt that 'measure the temperature change' was not clear enough and if other students repeated their experiment they may do this differently.

2 Why is it important that a plan can be followed by other students?

Elsa suggested they set a time to record the final temperature, say 3 minutes, but Jonathon disagreed as one of the reactions may be faster than the other.

3 Why do you think it is a good idea to record the highest temperature reached and not to measure at a set time?

They agreed on recording the highest temperature reached and so completed their experiment. Here are the results:

Acid	Temperature (°C)						
	Start 1	End 1	Change 1	Start 2	End 2	Change 2	Average change
hydrochloric	22.03	28.54	6.51	22.04	28.93	6.89	6.70
sulfuric	22.02	33.68	11.66	22.00	32.86	10.86	11.26

Their conclusion is that the type of acid does affect the temperature change. Sulfuric acid gave almost twice the temperature rise that hydrochloric acid gave.

4 Explain which set of data is the most precise.
5 The theoretical value provided by their teacher was that hydrochloric acid and magnesium should have increased by 6 °C and sulfuric acid and magnesium by 12 °C. Do you think their data is accurate? Explain why.

Just as the alphabet forms words, so chemical elements can combine and create a huge range of different chemical substances. When different types of atoms are chemically combined we describe the substance as a chemical compound.

● **Figure 1** Magnesium burning in oxygen. How could we get millions of different chemical substances from only about 90 different elements?

→ Compounds

Compounds are made from more than one **element** chemically joined together. Compounds are formed when two or more substances react together. When a new substance is formed, this is a chemical change.

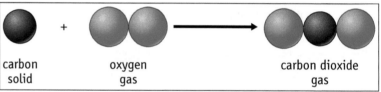

carbon solid + oxygen gas → carbon dioxide gas

● **Figure 2** When carbon reacts with oxygen a new compound, carbon dioxide is formed

Unlike the physical changes that occur when a substance changes state, chemical changes are much more difficult to reverse. This is because the atoms have rearranged in a new way. To reverse this, bonds that chemically combine the atoms would have to break.

Some compounds can be separated into their different elements by using other chemical reactions. Different types of chemical reactions have to be used, depending on the elements that the compound contains.

A particular compound always contains the same set of elements combined in the same way. Some examples of compounds include:

- sodium chloride
- copper sulfate
- carbon dioxide
- hydrochloric acid
- sodium hydroxide
- water.

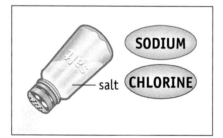

salt — SODIUM / CHLORINE

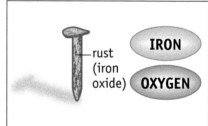

rust (iron oxide) — IRON / OXYGEN

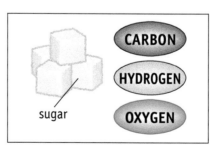

sugar — CARBON / HYDROGEN / OXYGEN

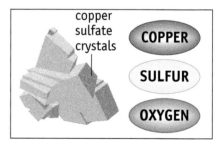

copper sulfate crystals — COPPER / SULFUR / OXYGEN

● **Figure 3** Examples of chemical compounds, and the names of the elements that combine to make them

→ Molecules

A **molecule** is the smallest part of a compound that has the properties of that compound. Molecules are made from **atoms** joined together by chemical bonds. All the molecules of a particular compound contain the same combination of atoms. Different compounds have different molecules, with different combinations of atoms.

→ Properties of compounds

When a compound is formed, it has its own properties. These are not necessarily the same as, or even a combination of, the properties found when the atoms existed as elements.

Take for example water, H_2O.

- melting point 0 °C
- boiling point 100 °C
- liquid at room temperature

These properties of water differ from the properties of the two gaseous substances that form if the atoms combine as the elements hydrogen (H_2) and oxygen (O_2).

Another example is copper sulfate; this contains atoms of copper, sulfur and oxygen.

- Copper sulfate is a bright blue crystalline substance.
- Copper is an orangey/brown metal.
- Sulfur is a yellow non-metallic solid.
- Oxygen is a colourless gas.

Compounds can be made when chemicals react together; this may happen in a laboratory, in a kitchen or inside a living organism. The right conditions are needed for the atoms to break apart and recombine in the new way. They often need an input of energy provided by heat.

● **Figure 4** Molecules in the compound water

● **Figure 5 Copper sulfate crystals**

Show you can...

Complete this task to show that you understand how compounds have differing properties from the elements from which they can be made.

Consider the compound sodium chloride. The elements that can make this are sodium and chlorine. Identify which of the three substances match each property:

- a white solid
- a yellow gas
- a silvery metal
- a non-metal
- a salt
- a poisonous gas
- a substance added to food.

? Questions

1 Which types of atoms does the compound copper sulfate contain?
2 Why are chemical changes more difficult to reverse than physical changes?
3 Describe the differences between a molecule of oxygen (O_2) and a molecule of water (H_2O).
4 Why do you think some reactions can happen at room temperature and others needs temperatures of 2000 °C to start?

It is essential for chemists to understand what makes up a chemical compound. Which atoms combine, how many different types, and in what numbers they are present, allows chemists to predict both the physical and chemical properties of a compound.

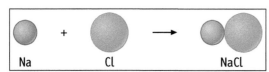

● **Figure 1** How many of each atom are present in this model of a glucose molecule?

→ Chemical formulae

The name of a compound usually tells you some of the elements it contains. Every compound has a chemical formula. Using the symbols for each atom as shown on the **Periodic Table** (page 83), the chemical formula tells you the atoms it is made of.

- sodium chloride, NaCl
- copper sulfate, $CuSO_4$
- carbon dioxide, CO_2

- carbon monoxide, CO
- hydrochloric acid, HCl
- water, H_2O

Though both carbon dioxide and carbon monoxide are made from carbon and oxygen only, they are different compounds because their molecules contain different proportions of oxygen.

To show the number of atoms in a formula a small number is placed immediately after the atom and is written below the line of the text (subscript).

- A sodium chloride molecule is made up of one sodium atom and one chlorine atom bonded together.

● **Figure 2 Representing sodium chloride**

- A water molecule is made up of two hydrogen atoms and one oxygen atom bonded together.

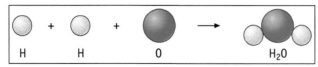

● **Figure 3 Representing water**

- A carbon dioxide molecule is made up of one carbon atom and two oxygen atoms bonded together.

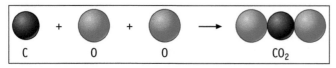

● **Figure 4 Representing carbon dioxide**

➜ Forming compounds

Compounds can form in three different ways.

1 A combination of metal and non-metal atoms:
   - the name of the metal comes first
   - the name of the non-metal follows and its ending is changed to –ide.

   For example: sodium and chlorine combine to make … sodium chloride.
   Magnesium and oxygen combine to make … magnesium oxide.

2 A combination of metal atoms, non-metal atoms and oxygen atoms:
   - the name of the non-metal part usually changes to include the ending –ate.

   For example: potassium, phosphorous and oxygen combine to make… potassium phosphate.

Non-metal	With oxygen
nitrogen	nitrate
sulfur	sulfate
carbon	carbonate
phosphorous	phosphate

3 Compounds formed between different non-metal atoms: some of these compounds do not have names that relate to their atoms, for example ammonia and methane. These compounds need to be learnt so that we can recognise them.

Name	Formula
water	H_2O
ammonia	NH_3
methane	CH_4

Notice that when hydrogen atoms are included in a compound the names and formulae do not follow these patterns. Remember your work on acids and alkalis and review the compound names and formulae below.

Acid	Formula	Alkali	Formula
hydrochloric acid	HCl	sodium hydroxide	NaOH
sulfuric acid	H_2SO_4	calcium hydroxide	$Ca(OH)_2$
nitric acid	HNO_3		

? Questions

1 What are the formulae of the following compounds?
   a) hydrochloric acid   b) water   c) carbon monoxide

2 What would you call the compound that forms between calcium and oxygen atoms?

3 What are the names of the different types of atoms in the following compounds?
   a) calcium nitrate
   b) potassium carbonate
   c) sulfuric acid
   d) sodium bromide

4 For sodium carbonate, Na_2CO_3, explain what the formula means in terms of the names of the different types of atoms and how many of each are in the compound.

5 Look at the compounds and their formulae in the table below. Look each atom up on the Periodic Table and describe any patterns that you notice, then predict the formulae for potassium oxide and calcium bromide.

Name	Formula
sodium chloride	NaCl
magnesium chloride	$MgCl_2$
sodium oxide	Na_2O
magnesium oxide	MgO

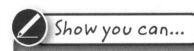 Show you can…

Complete this task to show that you understand how chemical compounds can be named and how to write their formulae.

Create some extra summary diagrams similar to those in Figures 2–4 for the following compounds:

- carbon monoxide
- hydrochloric acid
- calcium chloride ($CaCl_2$)
- ammonia.

6.3 Mixtures and compounds

To tell the difference between a mixture and a compound you would try and separate their components. Once atoms are chemically combined or bonded together it takes a chemical reaction to separate them. Mixtures on the other hand can be more easily separated.

● Figure 1 Mixture or compound?

→ Mixtures

You have a **mixture** when two or more substances are mixed together without any chemical reaction taking place. These substances could be elements or compounds. The substances in a mixture are not chemically bonded together.

Air is a mixture. Air contains several different gases, including:

- nitrogen
- oxygen
- carbon dioxide
- water vapour.

These gases stay separate from each other; they do not normally react to form new compounds. However, under the right conditions these elements can be combined.

The nitrogen and the oxygen in the air can combine to form the compound nitrous oxide. Nitrous oxide (N_2O) is sometimes called 'laughing gas' and was used by doctors and dentists for many years to relax patients.

A very similar reaction occurs inside car engines where the temperature is very high. The product in this case is nitrogen dioxide (NO_2), which is a major air pollutant leading to acid rain and smog.

Salt water is a mixture. Though the salt is dissolved in the water, the salt molecules do not chemically combine with the water molecules. They just fill up the spaces between water molecules.

Mixtures can be separated easily by physical methods (no chemical reaction is needed). The range of methods used to separate the substances in mixtures is included in Topic 4, pages 108–111.

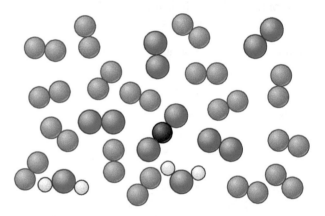

● Figure 2 Air is a mixture of compounds and elements, including the compounds carbon dioxide and water vapour, and the elements nitrogen and oxygen

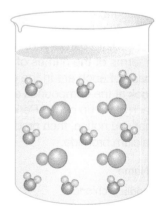

● Figure 3 Salt and water, mixed together but not chemically combined

→ Compounds

In a compound, the different substances it contains have chemically reacted together. The atoms of different elements in the compound are chemically bonded together. In order to separate a compound into its elements a chemical reaction is required.

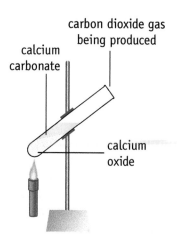

● **Figure 5** You can use heat to separate compounds by **thermal decomposition**

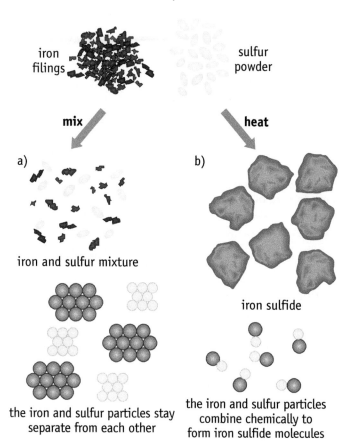

mix

a)

iron and sulfur mixture

the iron and sulfur particles stay separate from each other

heat

b)

iron sulfide

the iron and sulfur particles combine chemically to form iron sulfide molecules

● **Figure 4** The difference between a) a mixture of iron and sulfur and b) the compound iron sulfide

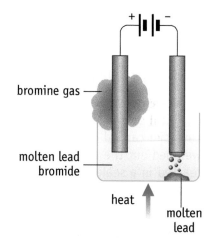

● **Figure 6** You can use **electrolysis** to separate some compounds. This works well for separating metals and non-metals

? Questions

1 Give the names and formulae for two of the compounds found in air.
2 What is the difference between a mixture and a compound?
3 Why will a magnet separate a mixture of iron and sulfur but will have no effect on iron sulfate?
4 Why do you think heat could be required to separate compounds?
5 Why do you think the lead bromide needs to be molten for the process of electrolysis?

✎ Show you can...

Complete this task to show that you understand the differences between mixtures and compounds.

Read each comment carefully — they all contain mistakes. Write out a correct version of each statement.

• Seawater is a compound — it contains sodium chloride combined with water.
• Water is a mixture — it contains hydrogen and oxygen. Its formula is H_2O.
• Oxygen is a compound. It is made up of two oxygen atoms combined together in a molecule of O_2.
• Sulfuric acid is a mixture because it contains lots of different atoms. Its formula is H_2SO_4.

6.4 Conservation of matter

Everything on the planet is made of matter. There is a scientific law that states that matter cannot be destroyed or created, only changed from one form to another. This means that over time atoms will be rearranged again and again yet the types and number of atoms remains conserved (unchanged).

● **Figure 1** How can it be true that all the matter on the Earth today was there when Earth formed billions of years ago?

→ Early ideas

As ideas about chemistry developed so did a curiosity about how chemical substances behaved during reactions. In 1774 Antoine Lavoisier carried out some very accurate experiments. He carefully weighed out the chemicals he wanted to combine, ensured they were reacted in a sealed glass vessel, and then carefully weighed all the products. He showed by repeated experimentation that the mass remained unchanged before and after a chemical reaction.

It was John Dalton's work on the atomic structure in the early 1800s (see Topic 2 page 82) that began to explain Lavoisier's observations. You will recall that every type of atom has its own **relative atomic mass**, based on the number of **protons** and **neutrons** found within its nucleus (see Topic 2 page 85).

→ Conserving mass

As molecules are two or more atoms chemically combined, it means that a molecule will also have a relative mass. If we take, for example, a molecule of oxygen, O_2, each oxygen atom has a relative mass of 16 and so the molecule will have a **relative formula mass** of $16 \times 2 = 32$.

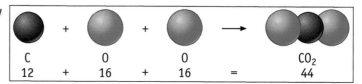

● **Figure 2** The relative formula mass of carbon dioxide

Since atoms must be present in the reactants to exist in the products, it stands to reason that overall mass will be conserved.

Conservation of mass is not only seen in chemical changes but also the physical process of dissolving. The mass of the solute, added to the mass of the solvent is then seen in the total mass of the solution (see Topic 4 page 104).

→ Challenging problems

The challenge with confirming the conservation of mass in a chemical reaction is the practical aspect of creating a sealed vessel. This can be difficult to achieve and can lead to experimental results which do not seem to support the law. The most difficult situations arise when a gas is either a reactant or a product, as determining the mass of a gas is somewhat challenging.

Combustion reactions

Take for example a combustion reaction. A piece of magnesium ribbon is carefully weighed and the mass recorded accurately as 0.024 g. The piece of metal is ignited in a Bunsen flame and the product is carefully collected. The mass is found to be 0.040 g and so the law of conservation of matter seems to be incorrect.

Consider carefully what has happened during the reaction. The magnesium ribbon is surrounded by air. During the reaction the oxygen molecules in the air have been rearranged with the magnesium atoms.

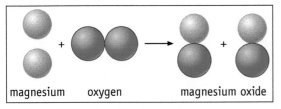

magnesium   oxygen   magnesium oxide

● **Figure 3** **The reaction between magnesium and oxygen during combustion**

The additional mass in the product is from the oxygen molecules. The difficulty is that there was no way to determine the mass of oxygen that reacted with the magnesium.

Similar problems arise if one of the products is a gas. The final mass appears to reduce when in fact some of the atoms have rearranged as a gas and left the vessel, leaving their mass unaccounted for.

❓ Questions

1 Which two subatomic particles make up the mass of an atom?
2 If 64 g copper reacts with 16 g oxygen, what mass of copper oxide will be formed?
3 Calculate the relative formula masses of the following compounds, given the following atomic masses: H = 1, O = 16, Ca = 40, N =14.
  a) H_2O
  b) CaO
  c) NH_3
4 Why was it so important for Lavoisier's experiments to take place in a sealed vessel?
5 What mass of oxygen reacted with the 0.024 g of magnesium in the example in the text above?

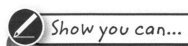

Show you can...

Complete this task to show that you understand the law of conservation of matter.

Explain the following observations.

1 When wood burns the mass of the ash is much less than the mass of the wood.
2 As I add sugar to water the volume stays the same but the mass keeps going up.
3 When I mix two chemicals together, after sealing the flask I can see a colour change but the mass remains the same.

Presenting and interpreting data

→ Mass matters

Two students, Joe and Jack, were asked to investigate the **law of conservation of mass**. Your challenge is to follow their method and review their data. You will need to identify patterns, use observations, measurements and data to draw conclusions.

Joe and Jack chose to use the reaction between calcium carbonate and hydrochloric acid.

calcium carbonate + hydrochloric acid → calcium chloride + water + carbon dioxide

For their first experiment they decided to find the mass of the starting substances and measure the mass at the end of the reaction.

They added the solid to the acid, and placed the tub that contained the carbonate back on the balance alongside the reaction mixture.

1 Why did they put everything on the balance together?

Their **hypothesis** was that the mass at the start would be equal to the mass at the end due to the law of conservation of matter. Here are their results:

Reaction	Mass at start (g)	Mass at end (g)
calcium carbonate + hydrochloric acid	127.03	126.48

2 Copy the table and add a column for mass change. Do not forget to include the correct units.

Their teacher reminded them that it is important to also think about your observations during an experiment and not just measurements. She asked them what they had noticed during the reaction.

3 This reaction produces a gas – describe what Jack and Joe would have seen?

They realised that the feature they had used to judge when the reaction had finished was the cause of the loss of mass.

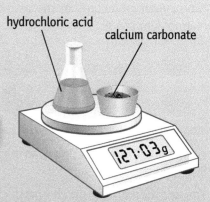

hydrochloric acid

calcium carbonate

127·03 g

● **Figure 1** Finding the mass at the start

Joe and Jack needed a way to collect the gas being produced in the reaction so they could find its mass. Joe did some research and found this diagram (Figure 2) on the internet.

Jack wanted to find out how the mass changed throughout the reaction, rather than just from the start to the end. They agreed a new plan:

- find the start mass of all flasks, chemicals and the balloon
- add the carbonate to the acid, start stopwatch and put the balloon over the top of the flask to collect the gas produced
- record the mass every 30 seconds
- repeat again with the same start mass to check for reliability.

Here are their results:

● **Figure 2** Using a balloon to collect gas produced in a reaction

Time (s)	Observations			Mass (g)			
	Try 1	Try 2	Try 3	Try 1	Try 2	Try 3	Average
0	nothing	nothing	nothing	140.00	140.00	140.00	140.00
30	lots of bubbles, balloon looks empty	lots of bubbles, balloon fell on floor	lots of bubbles, balloon looks empty	139.73	139.65	139.79	139.76
60	lots of bubbles, balloon starting to expand	Lots of bubbles, balloon just on and looks empty	lots of bubbles, balloon starting to expand	139.71	139.35	139.75	139.73
90	lots of bubbles, balloon still expanding	lots of bubbles, balloon slipped off again	lots of bubbles, balloon still expanding	139.71	139.40	139.75	139.73
120	fewer bubbles, no change in balloon	fewer bubbles, balloon looks empty	fewer bubbles, no change in balloon	139.71	139.45	139.75	139.73
150	no bubbles, no change in balloon	no bubbles, balloon empty	no bubbles, no change in balloon	139.71	139.45	139.75	139.73
180	no bubbles, no change in balloon	no bubbles, balloon empty	no bubbles, no change in balloon	139.71	139.45	139.75	139.73

4 Why is the start mass at time 0 so much higher than their first experiment?

5 What happened in try 2?

6 Why did they repeat a third time and identify try 2 results as outliers?

7 Why would the mass drop at the beginning of try 1 and try 3 when they had attached the balloon to collect the gas?

8 Comparing both the observations and the data for time 60 s to 180 s, describe how it supports the law of conservation of mass.

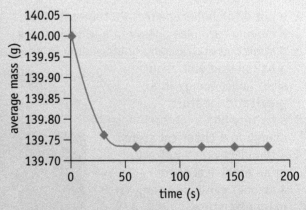

● **Figure 3** A graph to show change in average mass over time

As a rollercoaster climbs to the highest point on its track, it transfers energy from its electrical supply to a gravitational store. It then rapidly shifts this energy to a kinetic store as it rolls downhill.

● **Figure 1** Is a rollercoaster more exciting when its energy is in the gravitational or the kinetic store?

→ What is energy?

It is easier to talk about what you can do with **energy** than to define what energy actually is. Just as money gives the ability to buy things, energy gives the ability to do things. Energy gives a car the ability to move, a kettle the ability to heat up water, and a light bulb the ability to shine.

→ Energy stores

Your money could be stored in a savings account or in your pocket, or it could even be placed in a shop's till if you use it to buy something.

Energy can also be found in different stores. There are eight types of **energy store**:

1 a hot drink holds energy in a **thermal store**
2 a moving car holds energy in a **kinetic store**
3 a stretched elastic band holds energy in an **elastic store**
4 a ball placed on a high shelf holds energy in a **gravitational store**
5 a battery and a log both hold energy in a **chemical store**
6 a thunderstorm holds energy in an **electrical store**
7 a magnet can hold energy in a **magnetic store**
8 a radioactive atom holds energy in a **nuclear store**.

● **Figure 2** What kind of energy is stored in a moving car?

→ Shifting energy

Just as you transfer money from your pocket to a shopkeeper's till to buy something, energy can be shifted between the eight stores to do useful work.

- The elastic band's energy can be shifted to a kinetic store, when the rubber band is in flight, by **mechanical work**.
- The log's energy can be shifted to a thermal store in the surroundings by burning it. The energy transfer occurs by **convection (particle movement)** and by **radiation** of heat and light.
- The battery's energy can be shifted to a kinetic store in a spinning motor by **electrical work**.

→ Dissipating energy

Energy stores can be concentrated, like a kettle of boiling water. If you use this kettle of boiling water to try warm up a cold paddling pool, you have the same amount of energy in a more dilute form.

If you burn a log in a fireplace, the shift of energy to the thermal store of the surroundings is useful – it warms up your house. Often, though, when energy ends up in a thermal store, it is not useful at all. When a motor spins, the moving parts rub against each other, producing heat by friction. Here mechanical work transfers useful energy to useless energy in a thermal store: a hot motor.

→ Conservation of energy

The **law of conservation of energy** states that we can neither make, nor destroy, energy. All we can do is shift it from one store to another. When we have taken all of the energy from the chemical store of the wood, and transferred it to the thermal store of the surroundings, we have the same amount of energy as before, just stored in a different way.

● **Figure 3** This pile of money represents the energy stored in the kettle of boiling water. How would you represent the energy stored in the paddling pool?

 Questions

1 Give an example of a situation in which energy is held in:
   a) a gravitational store
   b) an elastic store
   c) a kinetic store.
2 A bow and arrow shift energy between stores. Can you name the energy stores, and give an explanation for the shift?
3 By what process is energy shifted:
   a) from the gravitational store of a rollercoaster at the top of its track, to the kinetic store of the rollercoaster when it is moving fast at the bottom of the track
   b) from the thermal store of the Sun to the thermal store of our environment here on Earth
   c) from the chemical store of a battery to the thermal store of an electric heater
   d) from the thermal store of a hot convection heater to the thermal store of the surrounding air?

 Show you can...

Complete this task to show that you understand energy stores.

Write the names of the energy stores involved in:

a) a ball rolling along the ground
b) a kettle of hot water
c) the water in a reservoir high up in the mountains
d) the food you eat
e) a stretched catapult.

1.2 Heat transfer and the thermal store of energy

People use ice to cool their drinks down, but the ice does not 'give the drink its coldness' – rather, the drink loses some energy from its thermal store to the thermal store of the ice, and so the drink cools and the ice melts.

● **Figure 1** Energy from the thermal store of the drink will melt the ice, but what happens to the total amount of energy in the glass?

→ The thermal store

The molecules in a substance are always moving, to a greater or lesser degree. In a warm liquid, the molecules move faster than in a cold solid. When the fast molecules bump against the slower ones, they give them some of their energy – the slower molecules speed up, while the faster ones slow down.

This is how a substance holds energy in a **thermal store**, and how that energy is passed on to another substance.

→ Warming and cooling

When a snowman melts on a warm winter's day, the snowman is warming up, but the heat from the air takes energy from the air's thermal store, so the air cools down.

Holding a hot mug of tea warms your hands because heat from the mug transfers energy to your hands. The tea also cools down.

After a race, marathon runners wrap themselves in 'space blankets' to reduce heat transfer, so that they don't cool down (and warm up the air around them) too quickly.

● **Figure 2** The particles in a warm substance (top) move faster than those in a cold substance (bottom)

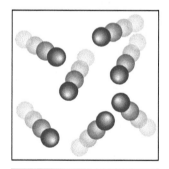

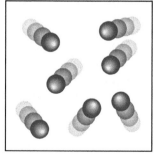

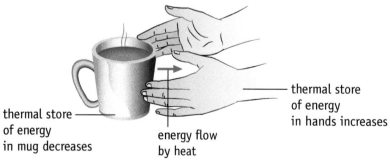

thermal store of energy in mug decreases

energy flow by heat

thermal store of energy in hands increases

● **Figure 3** The red arrow shows how energy is transferred from the mug's thermal store to your hands

→ Heating and cooling

Hot chocolate in a mug gradually cools down. It does this because it is losing energy. The energy transfers out of the hot liquid and into the surrounding room.

Energy can transfer between objects when one object is hotter than the other – so there must be a difference in temperature. In this case, the chocolate is hotter than the room. This is called **heat transfer**.

A drink of cold milk would gradually warm up if left in a warm room. This time the room is the hotter object, so the heat would transfer the other way, from the room to the milk.

→ Investigating temperature over time

You could investigate how the temperature of a drink varies with time. As temperature and time are both **continuous variables** (they can have any value), you should draw a line graph. The shape of the graph then helps you to draw a conclusion, for example about the time it takes the chocolate to reach a safe temperature to drink.

When you carry out investigations you need to measure the variables as **accurately** as you can. This means that your measurements of temperature and time are close to the **true values**. Some measuring instruments are more accurate than others, but even a good-quality measuring instrument can give inaccurate results if you make an error when you use it.

- Choose the best quality instrument available.
- Choose a **thermometer** with an appropriate range for your measurement.
- Remember to 'zero' the **stopclock** before using it.
- Remember to read the unnumbered marks in between the numbers that are marked on the thermometer's scale.

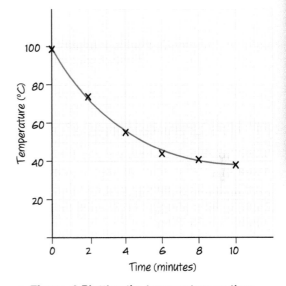

● **Figure 4** Plotting the temperature as time progresses clearly shows how a hot drink cools down

? Questions

1. When ice is put into a glass of cola, how does the thermal store of energy change:
   a) in the ice
   b) in the cola?
2. In question 1, what happens to the total thermal store of energy contained in the glass?
3. Draw an energy flow diagram, similar to Figure 3, to show how energy transfers from a marathon runner to the environment.
4. Draw an energy flow diagram to show the energy transfer when a snowman melts.

Show you can...

Complete this task to show that you understand heat transfer and the thermal store of energy.

Draw a diagram to show how the particles move in a gas with:

a) a large thermal store of energy
b) a small thermal store of energy.

The gravitational store of energy and work done

In October 2012 Felix Baumgartner jumped from a helium balloon 39 km above the Earth's surface and fell faster than the speed of sound. At 1360 km/h his kinetic store of energy was immense, but this energy must have come from somewhere.

● **Figure 1** Where did the energy in Felix's kinetic store come from?

→ The gravitational store

When you climb a staircase you transfer energy from the **chemical store** of your muscles to your **gravitational store**. The higher you climb, the more energy you transfer. A lighter person would need to transfer less energy to climb the same height, while a heavier person would require more.

We can see that the energy transferred depends both on the **weight** of the person and the height climbed.

● **Figure 2** You need to know your height and weight to calculate the energy in your gravitational store

The amount of energy transferred to the gravitational store is equal to the weight of the person multiplied by the height the person has climbed.

energy transferred =	weight	×	height
(in **joules**, J)	(in **newtons**, N)		(in **metres**, m)

→ Work done

The weight of a person is a **force**, and the height they climb is a distance. We can write the equation for energy transferred in more general terms, to tell us the amount of energy expended, for example, when a shopping trolley is pushed along. In this case we call this energy the **work done**.

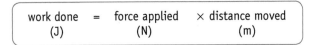

$$\text{work done (J)} = \text{force applied (N)} \times \text{distance moved (m)}$$

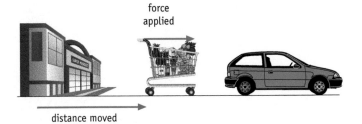

force applied

distance moved

● **Figure 3** The work done depends both on how hard and how far you push

→ Frictional heating

The smoke produced by this dragster is caused by the tyres spinning against the track and heating up due to **friction**. This heat burns the tyre rubber and causes smoke.

Friction is a very common cause of wasted energy. Whenever there are moving parts, friction transfers some of the energy of the moving parts to a thermal store. To make machines more efficient you need to reduce the friction.

● **Figure 4** Are there any situations in which the energy lost due to friction is useful?

→ Gravitational to kinetic

When a lift moves from the ground floor to the top floor of a block of flats, the work done by the lift motor is equal to the energy transferred to the lift's gravitational store. When it drops back to the ground floor, the motor does not need to do any work – energy simply moves from the gravitational store to the lift's **kinetic store**.

? Questions

1 Copy and complete the following sentences:
   a) When a skydiver falls from an aeroplane, energy is shifted from a .... store to a .... store.
   b) When a sledge is pulled along, energy is shifted from a .... store first to a .... store, and finally to a .... store.
2 Write down three situations in which the work done against friction causes a heating effect. In each case, say whether the heating is useful or wasteful.
3 A man weighing 800 N climbs 5 m up a vertical ladder. How much energy does he transfer to his gravitational store?
4 A cyclist pulls the brake levers and exerts a force of 100 N at the brake blocks. If she travels 4 m while stopping, how much energy has been transferred from her kinetic store to the thermal store of the environment?

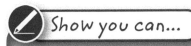

Show you can...

Complete this task to show that you understand work done and the gravitational store of energy.

Write a sentence to explain how the energy held in an object's gravitational store can be calculated by using the equation for work done.

1.4 Energy from fuels

Energy drinks are sold with the promise that they will enhance your performance, by giving you an 'energy boost'. But these effects are actually due to the caffeine in them, which in scientific terms contains hardly any energy at all.

● **Figure 1** Which ingredient of these drinks contains the most energy?

→ Energy from food

There are many chemicals in the foods that we eat, and our bodies need most of them, to a greater or lesser extent. But there are only three types of chemical that act as a store of energy, or **fuel**. These are **proteins**, **fats** and **carbohydrates** (starch or sugar). You can read about these in Biology Topic 6 page 64.

Look at the label on your breakfast cereal, or on the wrapper of your break-time snack. Does it mention energy? Is there a number given in **kilojoules** (kJ), or in **kilocalories** (kcal)?

→ Proteins, fats and carbohydrates

Our bodies break down the foods that we eat into their building blocks, which can then easily be used to increase our chemical energy stores. Proteins are used by the body to repair damage and to build new cells, and so are not used for energy as much as fats and carbohydrates. Our bodies are good at storing fat, and when we eat more than we need for energy, it is packed away underneath our skin 'for a rainy day'. Carbohydrates are used in **respiration**, or stored on a short-term basis in the liver.

It's rather like saving up for a new bike by putting money into a savings account (fat), while also keeping your pocket money in a moneybox, where you can dip into it from time to time (carbohydrate).

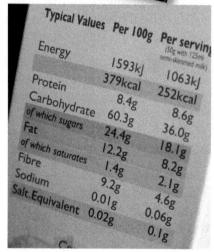

Typical Values	Per 100g	Per serving (50g with 125ml semi-skimmed milk)
Energy	1593kJ	1063kJ
	379kcal	252kcal
Protein	8.4g	8.6g
Carbohydrate	60.3g	36.0g
of which sugars	24.4g	18.1g
Fat	12.2g	8.2g
of which saturates	1.4g	2.1g
Fibre	9.2g	4.6g
Sodium	0.01g	0.06g
Salt Equivalent	0.02g	0.1g

● **Figure 2** What does the nutritional label on your breakfast cereal tell you?

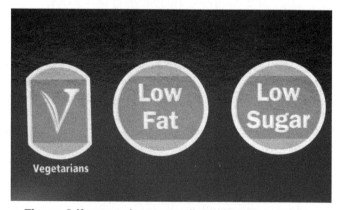

● **Figure 3** If we need energy to live, why are we told to eat a low-fat diet?

→ Comparing different fuels

You can compare the amount of energy stored in different fuels (including food – see Biology Topic 6 page 67) by using this apparatus. To make it a fair test you should use the same amount of fuel each time. You should also control other **variables** – use the same amount of water and the same starting temperature. Remember to wear eye protection.

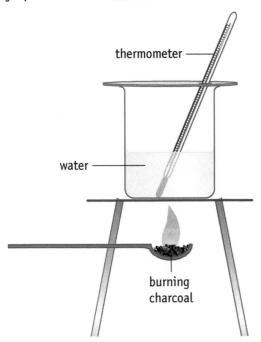

● **Figure 4** Burning transfers energy from the chemical store of the fuel to the thermal store of the water

The energy released from burning the fuel increases the temperature of the water. The fuel that releases the most stored energy will produce the largest temperature rise.

Unfortunately, lots of heat will be **dissipated** (lost) into the surroundings and so not all the heat is transferred to the water. This means the temperature rise may not be the true value. There could also be errors if the thermometer is not very accurate.

 Questions

1 Which chemicals in the food you eat act as fuels (stores of energy)?
2 What other fuels can you think of? (Not ones that you would eat!)
3 A grown man needs around 10 000 kJ of energy per day. One gram of carbohydrate provides 16 kJ, one gram of protein provides 17 kJ, and one gram of fat provides 37 kJ. Use these numbers to explain why a man eating a high-fat diet is more likely to put on weight than a man on a low-fat diet.
4 Nyan measured the temperature rise of two identical beakers of water when he burnt a peanut under one and a piece of popcorn under the other. Both foods weighed the same amount, but the water above the peanut got hotter than the water above the popcorn.
   a) What does this tell you about the peanut and the popcorn?
   b) Which food do you think contained more fat?

 Show you can...

Complete this task to show that you understand fuels and energy.

Draw an energy transfer diagram for the experiment shown in Figure 4

141

Power tools are very useful when doing gardening or DIY. A hedge trimmer will cut faster than a pair of shears, an electric drill works faster than a hand drill, and many people even have an electric screwdriver to help save time and effort.

● **Figure 1** What powers a lawnmower that does not run on electricity or petrol?

→ Power

All machines transfer energy from one store to another, but some do it faster than others. A more powerful kettle will heat up water faster than a less powerful one, and a more powerful car will have a greater acceleration than a less powerful one.

Power is the rate at which a machine does work, or the rate at which it shifts energy from one store to another.

● **Figure 2** The sports car has good acceleration, but does it have more power than the lorry?

→ Power of electrical appliances

Some electrical appliances have more power than others. The power of an electrical appliance is usually measured in **kilowatts** (kW). An appliance with a power of 1 kW (1000 W) will shift 1000 joules of energy in a single second.

Typical power ratings of appliances	
Appliance	Power (kW)
Kettle	2.0
Lawnmower	1.5
Microwave oven	1.0
Toaster	0.8
Television	0.2
Laptop computer	0.1

→ Energy efficiency

If you could design the perfect light bulb, it would transfer all the supplied energy as light. The **efficiency** of your perfect bulb would be 100%.

Unfortunately there are no devices that are 100% efficient – they all waste energy in some way. In almost all devices, the wasted energy is lost as heat. Modern low-energy light bulbs produce much less heat than older bulbs do. They are more efficient.

> total energy in (J) = useful energy out (J) + wasted energy out (J)
>
> efficiency (%) = $\frac{\text{useful energy out (J)}}{\text{total energy in (J)}} \times 100$

As energy is never created nor destroyed, you can work out the wasted energy if you know the input energy and the useful output energy. A light bulb with an input of 100J might transfer 20J of this to the surroundings as light. This mean it must transfer 80J as heat, as the total energy out (20J as light, 80J as heat) must equal the total energy in (100J). The efficiency of this light bulb (using the equation above) would be: $\frac{20}{100} \times 100 = 20\%$.

Modern **LED** bulbs are more efficient than 'energy-saving' **compact fluorescent bulbs**, which in turn are more efficient than old-fashioned incandescent **filament bulbs**.

● **Figure 3** If wasted energy is always lost as heat, could an electric heater be 100% efficient?

Questions

1 Which of the following is more powerful?
   a) a stereo system or a portable radio
   b) a mobile phone screen or a widescreen television
   c) a pocket torch or a light bulb
2 If you swapped the following machines for more powerful ones, what differences would you notice?
   a) a hairdryer
   b) a torch
   c) a leaf blower
   d) a toaster
3 The energy an appliance uses (in kJ) is equal to the power of the appliance (in kW) multiplied by the time it is used (in seconds). Using the power ratings given in the table on page 138 calculate how much energy is used when:
   a) a kettle is used for 2 minutes (120 seconds)
   b) a microwave oven is used for 10 minutes (600 seconds)
   c) a television is used for 3 hours (10800 seconds).
4 Calculate the efficiency of:
   a) an LED light bulb that transfers 95 joules usefully through light for every 100 joules it is supplied with
   b) a car that wastes 30 kJ through heat for every 40 kJ it releases from its fuel
   c) a lift that transfers 200 J to its gravitational store for every 800 J it transfers to the thermal store of the environment.

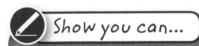

Show you can...

Complete this task to show that you understand power.

Write a sentence to explain why a television, with a power of 0.2 kW, will often use more energy in a day than a kettle, which has a power of 2 kW.

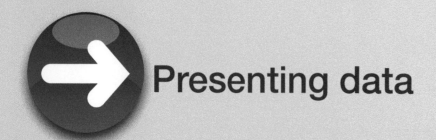

Presenting data

→ Television's hot!

Joel and Charlotte wanted to persuade their parents to buy a new television, so they looked up the energy performance of the leading models. They found out that, for the same sized screen, the LED television used 75 W, the LCD one used 110 W and the plasma screen type used 210 W. In each case, the television only gave out 5 W of this energy as light and 1 W as sound; the rest was transferred to the thermal store of the environment, as shown in this table.

● **Figure 1** When buying a television, most people don't look at the energy consumption. Why not?

	Television type		
	LED	LCD	Plasma
Light emitted	5 W	5 W	5 W
Sound emitted	1 W	1 W	1 W
Heat emitted	69 W	104 W	204 W
Total energy used	75 W	110 W	210 W

Presenting data

One of the skills scientists have to use is the ability to present their data in a meaningful manner, which helps the reader to understand the facts quickly and easily. Depending on what the data are, they could be presented as:

- a bar chart
- a pie chart
- a line graph
- a bubble chart
- a flow chart
- a **Sankey diagram**.

How to draw a Sankey diagram

A Sankey diagram is a bit like a flow chart, but one in which the width of each arrow indicates the amount of energy that is flowing into, or out of, the appliance.

In Figure 2, the incoming arrow is 10 squares wide, indicating 100% of the input power. The outgoing arrow on the right is only one square wide, which tells us that a filament bulb only transfers 10% (one square in 10) as useful light. Finally, the downward arrow represents all power that has been wasted – in this case it is nine squares wide, showing that 90% of the power is wasted as heat.

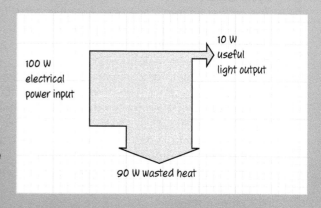

100 W electrical power input

10 W useful light output

90 W wasted heat

● **Figure 2** An old-fashioned filament light bulb is very inefficient

The task

Your task is to present the television data in these various forms, and to decide which would be the best way for Joel and Charlotte to present the information to their parents.

Remember that the energy used in 1 second by a 75 W appliance is 75 J. So your labels can be in either power (in watts) or energy used per second (in joules).

If you are working as a group, divide the following tasks among yourselves.

1 Draw a bar chart showing the energy emitted by the LED television.
2 Draw a pie chart showing the energy emitted by the LCD television.
3 Draw a line graph showing the energy emitted by the plasma television.
4 Draw a bubble chart to show the energy used and the energy emitted by the LED television.*
5 Draw a flow chart to show the energy input (total energy used) and the energy outputs of the LCD television.
6 Draw a Sankey diagram (see page 144) to show the energy flow through a plasma television.

* To draw a bubble chart for the LED television, draw circles whose sizes are in proportion to the numbers in the table, e.g. one large circle of 7.5 cm radius, a smaller one inside this of 6.9 cm radius, and two more of radii 5 cm and 1 cm. Colour code and label each bubble.

Comparing the diagrams

When your group has finished drawing their graphs, charts and diagrams, compare them.

7 Each member of the group should say why their chart is good for presenting the data.
8 After all group members have spoken, decide on which method of presenting this data you think Joel and Charlotte should use when talking to their parents

2.1 Force interactions

What is a force? The Royal Air Force is a fighting force, the police force can force open the door to a criminal's house, and the wind sometimes blows with gale force. Which of these is the most scientific use of the word?

● **Figure 1** What kind of force do you need to open a jam jar?

→ What is a force?

A **force** is a push or a pull. In everyday language, we might use the words 'tug' or 'yank' instead of pull, and perhaps 'shove' or 'hit' instead of push. In scientific language these are all forces.

→ What can a force do?

If you push a toy car, it will speed up. When you catch a ball, you slow it down and bring it to a stop. When you play with a yo-yo, you tug as it reaches the bottom of the string to make it change direction.

When you play with modelling clay your pushes and pulls make it change shape. The same happens if you stretch a rubber band, or sit on a space hopper – one stretches and the other squashes.

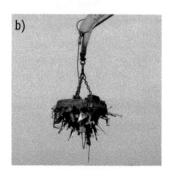

→ Common forces

You see many forces in everyday life:

- **applied forces** – pushes and pulls
- **gravity** – pulls a ball down when you throw it into the air
- **friction** – helps you to walk and stops you from slipping over
- **support force** – the chair you are sitting on pushes up on you to support your weight
- **magnetism** – pulls on steel objects such as paper clips
- **electrostatic force** – attracts dust to your television screen
- **upthrust** – enables boats to float
- **surface tension** – lets small insects walk on water
- **air resistance** – enables a parachutist to fall slowly
- **tension** – allows a rock climber to abseil safely down a rope.

● **Figure 2** What forces are acting in each of these pictures?

→ Contact and non-contact forces

You cannot throw a ball unless your hand actually touches it, but the Earth's gravity will pull it downwards even when it is in mid air. The push of your hand is a **contact force**, whereas the pull of gravity is a **non-contact force**.

→ Forces on an atomic level

If we took a really powerful microscope and looked at the ball, it would show us the individual atoms on its surface. When you pick up the ball, the atoms on the surface of your hand come close to the surface atoms of the ball. The outer layer of every atom is made up of negatively charged electrons, which repel (push away from) each other. The ball pushes on your fingers and your fingers push on the ball. So even contact forces are actually due to the non-contact electrostatic force between electrons.

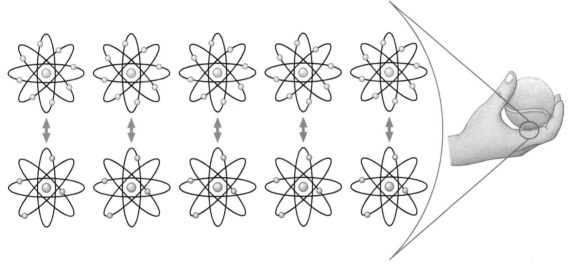

● **Figure 3** The outer electrons in the atoms of the ball (top row) repel the outer electrons in the atoms of your hand (bottom row)

? Questions

1 Look at Figure 2. List as many of the forces acting on the jet ski as you can think of.
2 Look at the list of common forces on page 146. Which ones are:
  a) contact forces
  b) non-contact forces?
3 Give examples of the following forces, and explain your reasoning:
  a) a non-contact force that can make something speed up
  b) a contact force that can change something's direction
  c) a non-contact force that can pull harder than gravity.

✎ Show you can...

Complete this task to show that you understand the different effects of forces.

Give an example of a force:

a) speeding an object up
b) slowing an object down
c) making a moving object change direction
d) changing the shape of an object.

Effect of forces on shape

If you stretch an elastic band and let go, it will return to its original length. This is an elastic deformation. If you squash a piece of Plasticine™, it will stay squashed after you let go. This is a plastic deformation.

● **Figure 1** How do the different properties of Plasticine™ and elastic bands suit them to their use?

→ Elastic or plastic?

If you sit on a car bonnet it will 'give' slightly under your weight, but it will return to its original shape when you get off. If the car is in a crash, the force of the impact, which is much larger than your weight, will permanently deform the bonnet.

● **Figure 2** Was this an elastic or a plastic deformation?

→ Stretching a spring

A spring stretches when you apply a force. The distance the spring stretches is called the **extension**. Here a **newton meter** is being used to measure the force of the **load** (**weight**) pulling down on the spring. Forces are measured in **newtons** (N).

Inside the newton meter is another spring. A larger force stretches the spring by a greater amount, and the marker moves further down the meter's scale. Reading where the marker is on the scale tells you the size of the force.

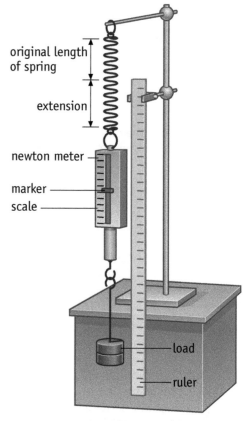

● **Figure 3** You can investigate how forces change the shape of a spring using this apparatus

148

Investigating extension of a spring

The extension of a spring depends on the force, so the extension is the **dependent variable** and the force the **independent variable**. The independent variable is the one that you change in an investigation – in this case by adding more masses to the load. To make it a fair test, you must keep all of the other variables (such as the type of spring) the same.

You should record all of your results in a table. It is a good idea to repeat your readings. Check that your repeat readings are similar. If they are not then you should take your measurements again. Do not forget to include the **units** in the top row of the table.

What safety precautions should you take when carrying out this investigation?

→ Hooke's law

Hooke's law states that the force exerted on a spring is **proportional** to its extension. The graph starts as a straight line through the **origin**, which shows that this law is true for the spring used. Doubling the force doubles the extension. This is no longer true beyond the elastic limit (see Figure 4), where the spring will not stretch in proportion to the force. Forces of this size will permanently change the shape of the spring.

When we stretch a spring, we do **work** on the spring to shift energy into its **elastic store**. The work done in elastic deformation is equal to half the stretching force multiplied by the extension. This energy can be released when we let it go.

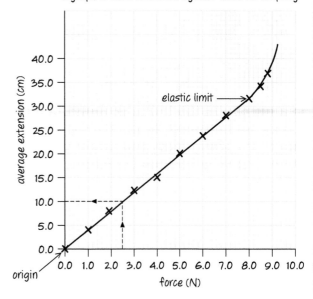

a graph to show extension against force for a spring

● **Figure 4** On a line graph you can read the extension for any force, even if it was not measured in the experiment

Show you can...

Complete this task to show that you understand the difference between elastic and plastic deformation.

Explain why a catapult would not work if the elastic band only deformed plastically, and why you could not make a model from modelling clay that only deformed elastically.

? Questions

1 Which part of the graph in Figure 4 represents:
   a) plastic deformation
   b) elastic deformation?

2 A watch spring, a clockwork toy and a catapult all store energy in the same way. In each case:
   a) name the energy store involved
   b) state what store the energy moves into once the object is released.

3 Use the graph in Figure 4 to answer the following questions.
   a) What force is needed to give the spring an extension of 25.0 cm?
   b) What is the spring's extension for a force of 4.5 N?

4 The following set of data was taken in an experiment on stretching a spring. Plot a graph of extension against force for this spring. Remember to give a title, and to label your axes fully. When you have plotted the points, draw a line of best fit. Does the spring obey Hooke's law?

Load (N)	Extension (cm)
0	0.0
2	2.5
4	5.0
6	7.5
8	10.0
10	16.0

2.3 Balanced and unbalanced forces

In a tug of war, the teams go nowhere if their forces are balanced. To win the tug of war, one of the teams must pull harder than the other, and so unbalance the forces.

● **Figure 1** If you knew the force of each person in the blue team, how would you find out the team's total force?

→ Balanced but moving

If the forces on an object are **balanced**, this does not always mean it is stationary. If the object is already moving, it will continue to move at a steady speed. The forwards force cancels out the backwards force, so it neither speeds up nor slows down.

The parachutist in Figure 2 is not hanging in mid air. He is falling to the ground at a steady speed, as the force of **air resistance** is balancing the force of **gravity**.

An **unbalanced force** is needed to make an object speed up or slow down, and to make it change direction.

→ Finding a resultant

Forces have a size and a direction. We can draw an arrow to represent a force. The longer the arrow, the bigger the force. The direction of the arrow shows the direction of the force.

To find the **resultant** (overall) force on an object, you simply add up all of the forces acting on it, calling the forwards forces positive and the backwards forces negative.

If the forwards forces equal the backwards forces, then we say that they are balanced. The resultant is zero.

— force of air resistance

force of gravity —

● **Figure 2** Air resistance balances the parachutist's weight

The effects of balanced and unbalanced forces

If the forces on an object are balanced (resultant = 0 N) then it will either stay stationary, or it will carry on moving at the same speed in the same direction, until an unbalanced force acts on it.

If the resultant is in the same direction as the motion, the object will speed up. If the resultant is in the opposite direction to the motion, the object will slow down. The directions that are positive and negative do not matter, as long as they are opposite to each other.

Support force

When a book sits on a desk, gravity pulls it downwards towards the ground. But the book does not move, so the forces on it must be balanced. The desk must be pushing back upwards with a force equal to the book's weight. This force has a number of names: the normal force, the reaction force and the **support force**.

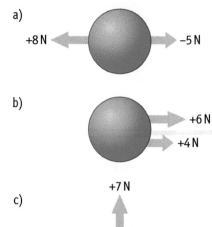

a)

b)

c)

● **Figure 3** What is the resultant force in each case here, and in which direction?

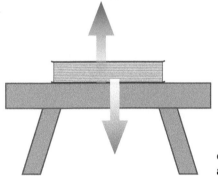

● **Figure 4** The support force is equal and opposite to the weight

? Questions

1 A book sits on a bookshelf. How can you tell that the forces on the book are balanced?

2 A parachutist falls through the sky at a steady 5 m/s.
   a) Are the forces on her balanced?
   b) How can you tell?

3 a) A car accelerates along the motorway. Is the resultant force forwards or backwards?
   b) What happens to the resultant force when the car reaches a constant speed?
   c) The car now starts to slow down. Is the resultant force now forwards or backwards?

4 What is the resultant force in each of these situations?
   a) A man pushes a box across the floor with a force of 140 N. The force of friction is 100 N.
   b) A freefall skydiver has a weight of 700 N. Just after she jumps from an aeroplane, the force of air resistance on her is 80 N.
   c) Two tug-of-war teams take the strain. The team on the right pulls with a combined force of 840 N. The team on the left only manages to pull with 780 N.

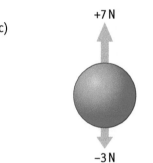

✎ Show you can...

To show that you have understood this chapter, copy out the following paragraph, filling in the blanks:

When the forces on an object are ...., it neither speeds up nor slows down, but continues to travel at a .... speed in a .... line, until an .... force acts upon it. An unbalanced force acting in the same direction as its motion will make it ...., whereas an unbalanced force acting in the opposite direction to its motion will cause it to .... To make it change direction also requires an .... force.

151

2.4 Friction forces

If you have ever tried roller-skating, you will know how much you need friction to get you moving. If you have roller-skated and tried to stop, you will know how important friction is in this situation too.

● **Figure 1** Roller-skates reduce friction, but how does this affect your ability to start and stop?

→ Friction can be a nuisance

Look at Figure 2. When Max tries to push the washing machine with a small force it stays still. The force of **friction** balances Max's pushing force. Max has to push really hard to make the washing machine slide along. His force needs to be larger than the maximum friction force. The forces are then unbalanced and the washing machine starts to move.

● **Figure 2** Max wants to push his new washing machine into the right place

Friction is a force that acts when two surfaces move against each other. This force either stops objects from sliding or, once they're moving, acts to slow them down.

→ What causes friction?

If you look closely at surfaces you can see that they are quite rough. In order for surfaces to slide past each other they need to lift up and down a little to get past all of the bumps. Therefore Max needs to apply a force so that his washing machine can slide along the floor.

There are things that Max could do to reduce the friction. He could make the surfaces as smooth as possible or use a **lubricant**. This gets between the bumpy surfaces and holds them slightly apart.

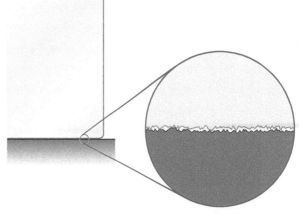

● **Figure 3** On a microscopic level, even smooth surfaces are bumpy

→ Friction can be useful

Friction can be a very useful force. High-tech climbing shoes have special rubber soles that increase friction between feet and rock. In a similar way, brake pads use friction to oppose the motion of a bicycle. Applying the brakes unbalances the forces and makes the bicycle slow down.

● **Figure 4** Why are the soles of climbing shoes made of very soft rubber?

? Questions

1 In each of the following situations, is friction a nuisance or useful?
   a) rubbing your hands together to keep them warm
   b) the moving parts of a car engine
   c) a slide in a park
   d) a pencil writing on a sheet of paper
2 How could Max reduce the friction under his washing machine to make it easier to move?
3 Would you be able to walk if there was no friction between your shoes and the ground?
4 Ice skates and running shoes both change the amount of friction between your feet and the ground. Find out whether they increase or reduce friction, how they do this, and why.
5 Lubricating a creaky door hinge with oil can silence it. Explain why the hinge creaked before it was lubricated, and why it was silent afterwards.

✎ Show you can...

Complete this task to show that you understand friction.

Explain why you could build a slide out of smooth plastic or polished metal, but not out of roughly sawn wood.

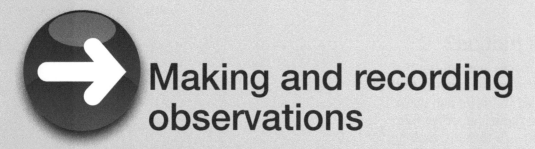

Making and recording observations

→ Slippery shoes

Have you ever slipped over on a wet floor? Or do you find some surfaces naturally more slippery than others? It is important for businesses such as hotels and shops to know which surfaces are slippery to avoid customers falling and hurting themselves.

The apparatus shown in Figure 1 can be used to investigate the friction between your shoes and different floor materials.

There is a force down the slope due to gravity. If the force of friction acting up the slope balances this, then the shoe is in **equilibrium** and will remain at rest.

If you make the slope steeper, the shoe eventually begins to slide. The force down the slope is now bigger than the maximum force of friction.

force of friction

shoe begins to slide when the forces become unbalanced

lab jack

force down slope increases as slope is made steeper

● **Figure 1** If there was no friction then this shoe would slide down the slope

You can investigate the maximum friction between different surfaces by measuring the height or angle of the slope when the shoe begins to slide.

A leading hotel chain investigated several different surfaces to see which one offered most grip and which was most slippery:

   a) laminate flooring
   b) stone tiles
   c) linoleum
   d) wooden decking
   e) deep pile carpet
   f) shallow pile carpet.

In the investigation, all of the other variables that affect friction need to be kept the same to make it a fair test. We call these **control variables**. In this experiment we need to use the same shoe throughout.

> **1** Can you predict the results that the hotel chain recorded?

2 If you have the equipment available, you could investigate the same surfaces. You could also try some of these surfaces when wet.

type of surface	height 1 (cm)	height 2 (cm)	height 3 (cm)	average height (cm)
laminate flooring	56	52	54	54
decking	50	52	51	51
decking (wet)	46	48	44	46

● **Figure 2** To make sure your results are reliable you need to repeat your readings

The hotel investigation showed the following results:

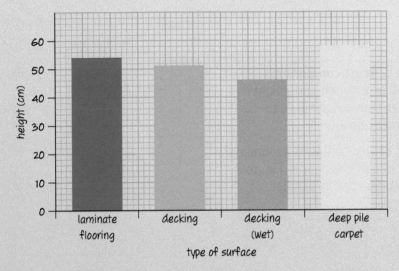

● **Figure 3** Think about how you will present your results

3 Results need to be presented in a way that is easy to understand and that makes the outcome of your experiment clear to see. Decide how best to present the results and draw up your table and/or graph.

4 Write a paragraph summing up the findings. Here are some points to think about:

a) Which surfaces are the most slippery?

b) Which surfaces are the least slippery?

c) Are the surfaces suited to their uses?

d) If not, what surface would you recommend to the hotel instead?

A microprocessor in a computer uses hundreds of millions of components. If you wanted to draw the circuit diagrams you would need a piece of paper that covers about four football pitches!

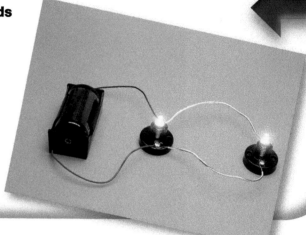

● **Figure 1** How would you draw a diagram for this circuit?

→ Electrical circuits

Even a simple electrical device needs several **circuit components**. A kettle contains an electrical heater as well as a switch and a light to show when it is on. A circuit diagram uses standard symbols to represent the different components in electrical circuits.

You can connect up components in **series** or in **parallel**. Components in series form a single loop, as in the top circuit diagram in Figure 2. The lower diagram shows two light bulbs connected in parallel.

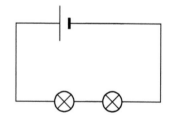

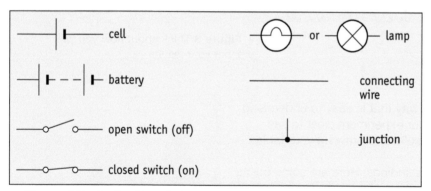

● **Figure 3** Basic circuit symbols

cell

battery

open switch (off)

closed switch (on)

or — lamp

connecting wire

junction

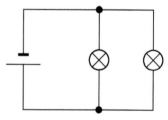

● **Figure 2** The bottom diagram represents the photo in Figure 1. When you know the symbols you can draw any circuit

If you put a switch anywhere in a series circuit, it can switch the whole circuit off. Switches in parallel branches work differently – each light can be switched on and off separately.

→ Electric current

In an electrical circuit, a battery or power supply pushes electrons through wires, from the negative side of the battery to the positive side (you will learn more about why electrons move in this direction in Topic 7 of Book 2).

In an ionic solution, negative ions (atoms that have gained an extra electron) will flow in the same direction that electrons would, whereas positive ions (atoms that have lost an electron) will flow in the opposite direction, from the positive electrode to the negative electrode.

The direction of the **electric current** (I) is defined as the direction that positive ions would flow in a solution – from the positive side of the battery to the negative side. Unfortunately, current was defined this way before the electron was discovered, when nobody knew what was actually flowing inside the wires.

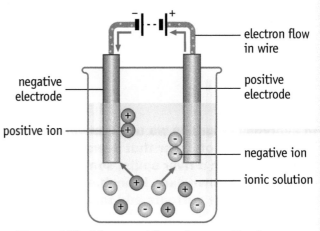

● **Figure 4** Electric current flows from positive to negative, the way that positive ions flow, in the opposite direction to electron flow

Labels: electron flow in wire; negative electrode; positive electrode; positive ion; negative ion; ionic solution

→ Potential difference (p.d.)

The **potential difference** (p.d.) of the **battery** or **cell** causes electrons or ions to flow in a complete circuit.

→ Measuring current and p.d.

The current flowing through a circuit is measured in **amperes** or **amps** (A) using an **ammeter**, which is placed in series (in line) with the circuit components. The p.d. is sometimes called the **voltage**, because it is measured in **volts** (V) using a **voltmeter**, which is placed in parallel (connected at either side of a component).

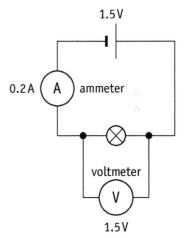

1.5 V · 0.2 A · A ammeter · voltmeter · V · 1.5 V

● **Figure 5** The ammeter is placed in series with the components, whereas the voltmeter is placed in parallel across a component

? Questions

1 a) Are the lights in your home wired in series or in parallel?
   b) How can you tell?
2 Find out the circuit symbols for:
   a) a heater
   b) a resistor
   c) a variable resistor.
3 Give an advantage that a parallel circuit has over a series circuit.
4 a) Copy the circuit in Figure 6, and label the switches and the bulbs.
   b) Which switch or switches would you need to close to light bulb 1?
   c) Which switch(es) would you need to close to light bulb 2?
   d) Which switch(es) would you need to close to light bulb 3?
   e) Which switch(es) would you need to close to light all of the bulbs?
   f) Draw in another bulb, which is not affected by any of the switches.

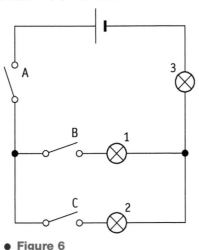

● **Figure 6**

✎ Show you can...

Complete this task to show that you understand how particles move around a circuit as an electric current.

List the particles that will flow from positive to negative in Figure 4, and those that will flow from negative to positive.

157

In everyday language we use the word 'current' for the volume of water that flows each second in a river. When a river splits, some of the current goes down each branch. When two rivers join, their currents add together.

● **Figure 1** How does the current flowing around the island compare to the current in the river before it split?

→ A model for current and voltage

We can use a model to compare current in a circuit with something that behaves in a similar way – a hot water system (see Figure 2):

- both must be complete circuits to work
- the **electric current** is like the flow of water
- the cell transfers energy to the current and pushes it around the circuit and through the light bulb – the boiler and pump heat the water and push it around the pipes and through the radiators
- the **potential difference** (p.d.) is similar to the temperature change of the water.

→ Rules for water circuits

In a **series circuit**, the amount of water that leaves the pump is the same as the amount that passes through each radiator, and the same as the amount that returns to the pump.

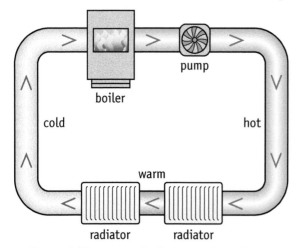

● **Figure 2** The current is the same throughout a series circuit…

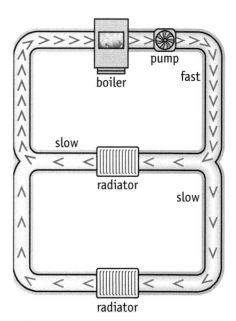

● **Figure 3** …but it splits and recombines at junctions in a parallel circuit

In a **parallel circuit**, some of the current can pass along each branch. The current that flows through the pump is equal to the currents in all of the branches added together.

→ Rules for electrical circuits

In a series circuit the p.d. of the cell is shared between the other components. If the two light bulbs are identical, then the p.d. across each bulb will be half of the p.d. of the cell.

Adding more light bulbs in series makes them dimmer because they have a smaller share of the p.d.

Light bulbs in parallel are all as bright as each other, because they each receive the full p.d. from the cell. Adding more bulbs does not make them dimmer, but it will make the cell run down faster.

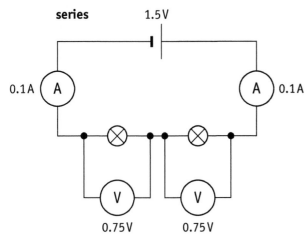

● **Figure 4** Current in a series circuit is constant, and the p.d.s across each bulb add up to the p.d. across the cell

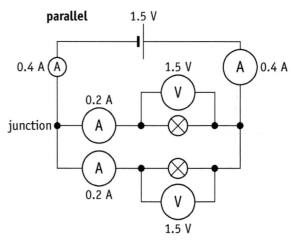

● **Figure 5** In a parallel circuit the current splits at a junction, and the p.d. across each component is equal to that across the cell

Questions

1 How does the current vary in a series circuit?
2 a) What is the missing current in this diagram?
   b) Is it entering or leaving the junction?
3 What is the rule for current at a junction?
4 Show how the hot water system model explains the p.d. in series and parallel circuits.
5 What is the rule for p.d.:
   a) in a series circuit
   b) in a parallel circuit?

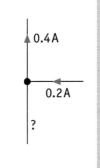

Show you can...

Complete this task to show that you understand this model for an electrical circuit.

List what is represented by each of the following:

a) the boiler and pump
b) a radiator
c) the temperature drop across a radiator
d) the current of water flowing through the pipes.

3.3 Resistance

Copper is an excellent electrical conductor, so much so that every cable in your house is made of it. But if you touch the copper, the current will flow through you. A plastic coating makes it safe, because plastic is an insulator.

● **Figure 1** Why does an electrician's screwdriver have a plastic handle?

→ Current depends on resistance

It is harder for electric current to flow through some objects than through others. These objects are poor conductors, and have a high electrical **resistance**. The greater the resistance, the smaller the current. If the resistance is very large, almost no current flows at all.

We can use the idea of a water circuit to help explain the effect of resistance. If a pipe is narrow, it will only let a small volume of water through each second (a small water current). We could say that it has a high resistance to the water current. If the pipe is wide, it will allow a large water current to flow. We could say that this pipe has a low resistance.

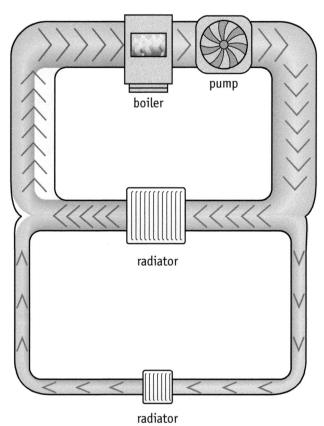

● **Figure 2** The wide pipe has a low resistance, and lets a larger current of water flow

→ Conductors and insulators

Good **conductors** are materials that have a low electrical resistance. Good **insulators** have such a high resistance that virtually no current will flow through them in normal use.

Generally speaking, metals are good conductors, and non-metals are good insulators. Two exceptions to this rule are graphite and water, which are both non-metals but can conduct electricity.

● **Figure 2** Why is the wire inside this cable made of copper, while the coating is made of plastic?

→ Current depends on potential difference

The size of a current also depends on the **potential difference** (p.d.) of the power supply.

A power supply with a low p.d. will push a small current through a conductor. Increasing the p.d. across the conductor increases the current that flows through it.

If you keep increasing the p.d. across an insulator, even this will eventually conduct a significant current.

For example, your skin is a good electrical insulator at low voltages, having a high resistance. If you touched the terminals of a 12 V car battery, a minute current of 1.2 mA would flow through your body – you would not even notice it. But if you touched the wires in a mains plug, at 230 V, then 23 mA would pass through your body. This would be enough to stop your heart.

? Questions

1 **a)** Why are wires generally made of copper, but their coatings are made of plastic or rubber?
   b) Why are light switches in bathrooms often operated by pull strings?
2 If the resistance of a circuit increases, what will happen to the current flowing through it?
3 A wire with a low resistance will let a larger current flow than a wire with a higher resistance would. How would you model these wires in a water circuit?
4 A bird can safely sit on an overhead electrical wire, but if you got a kite tangled up in one, you would be electrocuted. Why?

✎ Show you can...

Complete this task to show that you understand the concept of resistance.

Copy the following sentence, using the correct words from the brackets.

Electrical conductors have a (low/medium/high) resistance, and a (small/large) p.d. will make a (small/large) current flow through them. Insulators have a (low/medium/high) resistance, and even a (tiny/large) p.d. will not make a noticeable current flow through them.

3.4 Resistance calculations

A multi-lane motorway can take more traffic than a small single-lane road. The number of cars that pass along a road each second is another model for electric current, and the narrowness of the road represents the resistance.

● **Figure 1** In this model of electric current what could the moving cars represent?

→ Resistance, current and p.d.

The resistance of a wire or an electrical component is the ratio of the potential difference (p.d.) across it to the electric current flowing through it. Put another way, it is equal to the p.d. required to cause 1 A of current through the component. This can be summarised by the equation:

$$\text{potential difference} = \text{current} \times \text{resistance}$$
$$\text{p.d.} = I \times R$$

Resistance is measured in **ohms**, and the symbol for ohms is the Greek letter omega: Ω.

If the resistance of a wire were to double, then the current through it would halve – otherwise the p.d. across it would need to double to keep the current constant.

→ Changing resistance

Overhead electricity cables need to carry large currents. They are made of thick metal wires to make it as easy as possible for the current to flow. The electricians who work on these cables wear suits woven with silver wire, which conduct electricity very well. In this way, electric current from the cables runs through their suits, instead of through their bodies.

● **Figure 2** Do the electricians' suits have a low or a high resistance?

The metal filaments in light bulbs are very long and thin.

- The longer a wire is, the higher its resistance.
- The thinner a wire is, the higher its resistance.

A light bulb filament has a higher resistance than an overhead cable, so a smaller current flows.

● **Figure 3** How would you work out the current that flows through the filament?

→ p.d. and energy

Another way to think about p.d. is in terms of the difference in energy between two parts of a circuit. There is a p.d. (or voltage) across the light bulb because it is transferring energy to the environment through light and heat. You would measure the same p.d. across the cell. This is because the cell is transferring the same amount of energy from its chemical store as electricity.

? Questions

1 What are the units for the following (give the name and the symbol)?
   a) potential difference
   b) current
   c) resistance
2 If the p.d. across a circuit component increases, what will happen to the current flowing through it?
3 Rearrange the following equation:

   p.d. = I × R

   a) to show how you would find the current that flows through a component, if you knew p.d. and R
   b) to show how you would find the resistance of a component, if you knew p.d. and I.
4 Use this equation to answer the questions that follow:

   potential difference = current × resistance

   a) What p.d. is needed to make a current of 5 A flow through a wire that has a resistance of 2 Ω?
   b) What current would flow through a bulb which had a resistance of 100 Ω, when connected to a 9 V battery?
   c) What is the resistance of a motor that allows a current of 10 A to flow through it when connected to a 230 V supply?

Show you can...

Complete this task to show that you understand how the shape of a conductor can affect its resistance.

Copy the following sentence, using the correct words from the brackets.

A long wire will have a resistance (lower than/the same as/higher than) a short wire.

A thick wire will have a resistance (lower than/the same as/higher than) a thin wire.

Asking questions and making predictions

→ Lighting up the world

Who made the first light bulb?

Most people 'know' that American inventor Thomas Edison invented the electric light bulb. What many people do not realise is that British inventor Joseph Swan had already invented it a year earlier, in 1878. Before this, the world was lit by oil lamps, gas lights and candles.

The two inventors were inspired by Humphry Davy, who, in 1802, had connected a thin strip of platinum to an electrical supply. The platinum had become so hot that it glowed.

To turn Davy's laboratory experiment into a working electric light, Edison and Swan each had to develop a line of enquiry and use their scientific knowledge to plan a series of investigations.

They had to make observations during their experiments and then use their scientific understanding to work out what their observations told them.

They then asked further questions and made predictions, which they tested with more experiments.

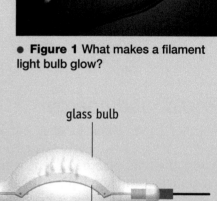

● **Figure 1** What makes a filament light bulb glow?

glass bulb

electric current in

electric current out

air, containing oxygen   thin platinum strip

● **Figure 2** Davy's light was not very bright, and the platinum strip soon burnt out in the oxygen in the bulb

Task 1

1 Imagine that you are Edison or Swan. What would you want to investigate as you tried to make the world's first working electric light? Make a list of the questions that you might ask.

* Think about the materials that you could investigate for the 'filament'. Which properties of these materials would you want to investigate?
* The platinum wire glowed because it became hot when electricity passed through it. The power output of the filament determines how bright the bulb is. The equation for the power output (P) is:

> power = current squared × resistance
> $P \quad = \quad I^2 \quad \times \quad R$

What questions might Edison or Swan have asked themselves when trying to find out how to increase the power output?

Table of melting points of conducting elements	
Element	**Melting point (°C)**
potassium	63
lead	328
magnesium	650
silver	961
gold	1063
copper	1084
nickel	1453
iron	1500
platinum	1770
tungsten	3400
carbon (graphite)	4000

Task 2

Now to develop your ideas into a line of enquiry. How would you proceed from here? What tests or experiments might you want to do? In what order? Remember that a scientist changes only one **independent variable** at a time.

1 If you were testing a wire made from one material, what tests might you do to try to make it brighter? List the tests that you would want to try.
2 If you were testing different materials, what variables would you have to keep constant to make it a fair test?
3 What else might you test for, as well as making the brightest filament?

Task 3

1 You would also have to prevent the filament from burning.
   a) What three things are needed for a fire? Research the 'fire triangle'.
   b) What could you try, to stop the filament burning? Think of the gas inside the glass bulb.
2 You know that the power of the bulb increases as either the **electric current** or the **resistance** increases (as long as the other stays constant). Use this knowledge to predict what would happen if:
   a) the filament was longer (thickness constant)
   b) the filament was thicker (length constant)
   c) the potential difference of the supply was increased.
3 One gas in particular reacts with hot materials to make them burn. Predict whether the wire would burn if the light bulb was filled with:
   a) air
   b) nitrogen
   c) oxygen
   d) carbon dioxide
   e) methane
   f) nothing – i.e. a vacuum.

Edison's and Swan's bulbs used carbon instead of platinum. Edison also realised that a very thin 'filament' with high electrical resistance would glow with only a small current, which made it much easier to supply electricity to the bulbs.

Task 4

1 Why do you think Davy used platinum in his original experiment?
2 What problem might you have selling light bulbs made with platinum?
3 Why do you think both Edison and Swan decided to use carbon for their filaments?

● **Figure 3** How could you stop the filament from burning out?

4.1 Energy from the Sun

Walk out of the shade on a sunny day and you begin to feel hot very quickly. Energy transfers from the thermal store of the Sun to your body. But be careful – some of the Sun's rays are harmful.

● **Figure 1** If there are no particles in the vacuum of space, are these people receiving energy by conduction, convection or radiation?

→ Solar radiation

The heat that we feel on a hot, sunny day is the Sun's **infrared radiation**. Infrared radiation is also responsible for the heat that we feel when we are beside a fire. **Light** is another way that energy reaches us from the Sun, and so is **ultraviolet radiation**. These are all forms of **electromagnetic radiation**.

→ Fossil fuels

Fossil fuels were formed very slowly from layers of dead plants and animals. They are a chemical store of energy. The ultimate source of this energy was the Sun.

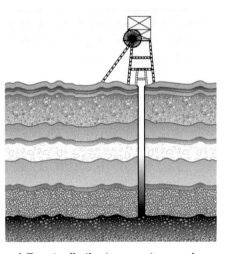

● **Figure 2 a)** Millions of years ago, plants absorbed energy from the Sun as they grew. Some energy passed up the food chain to animals

b) Dead plant and animal material collected at the bottom of swamps and oceans and was buried by layers of sediment

c) Eventually the temperature and pressure grew so great that fossil fuels formed. Today we extract these fuels as a valuable resource

Fossil fuels are **coal**, **crude oil** and **natural gas**. Coal formed over millions of years from plant material that became buried in mud in swamps. Oil and gas formed from the remains of sea creatures that became buried by sand under the ocean floor.

We use **fossil fuels** to heat our homes and to generate electricity. Crude oil can be separated into many different types of fuel such as petrol or diesel for cars and paraffin for jet aircraft.

→ Biomass

Plants absorb the Sun's energy and use it in photosynthesis. You can read about this in Biology Topic 5 page 54. This converts carbon dioxide and water into a chemical store of energy, which we call **biomass** (such as wood, leaves or seeds). We can use this energy resource by burning or eating the plant material. The Sun's energy passes up the food chain. We can even use biomass to generate electricity.

→ Global warming

Fossil fuels and biomass contain carbon. When they are burnt, the carbon is released into the atmosphere as carbon dioxide gas. Carbon dioxide traps heat in the atmosphere, warming the Earth up. It is feared that the increase in the Earth's temperature due to high levels of carbon dioxide – **global warming** – is starting to cause irreversible **climate change** (changes in long-term weather patterns).

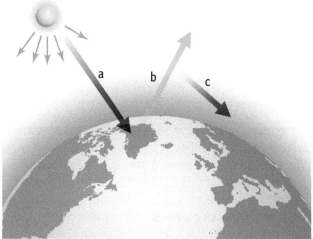

● **Figure 3** The Sun warms the Earth (a), which radiates heat into space (b). Carbon dioxide traps some heat (c), causing global warming

Questions

1 **a)** Name three forms of electromagnetic radiation that we receive from the Sun.
   b) Which of these is another name for radiated heat?
2 Where did the energy that is contained in a fossil fuel's chemical store originally come from?
3 Name a fossil fuel that formed from:
   a) dead plants
   b) dead animals.
4 The energy in biomass can usefully be released in two different ways. One of these ways is the same as for fossil fuels, the other is different.
   a) Name the process that can release energy from both biomass and fossil fuels.
   b) Give an example of this process for:
      i) biomass
      ii) fossil fuels.
   c) State the second way that energy can be released from biomass, but not from fossil fuels.

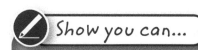

Show you can...

Complete this task to show that you understand that the Sun is the source of most of our energy here on Earth.

Explain how the energy in a bowl of breakfast cereal comes from the Sun.

In most power stations only 30 to 40% of the available energy is transferred usefully as electricity. The majority is lost through the large cooling towers of the power station. It is not smoke they give out, but clouds of water vapour.

● Figure 1 The clouds of water vapour carry away heat. How might this heat be used, rather than wasted?

→ Energy resources

An **energy resource** is something that stores energy in a form that we can easily use. Fossil fuels and biomass are energy resources. We can use their chemical stores to heat buildings, drive engines or generate electricity. Understanding different resources is very important. Some are quickly running out, or can cause environmental problems such as **global warming** or **acid rain**.

Energy resources are either **renewable** or **non-renewable**. Non-renewable resources – that is **fossil fuels** and **nuclear fuel** – will one day run out. Most of our electricity is generated from these non-renewable resources. Once we have used them they are gone forever.

→ Power from coal and oil

In coal and oil **power stations**, burning coal or oil shifts energy from the fuel's chemical store to the thermal store of water, which changes into steam. The energy transferred to the steam increases the kinetic energy of its particles. The faster these particles move, the more force they exert. The steam turns large fans called **turbines**. These turbines turn a **generator**, which produces electricity. At every stage some energy is **dissipated**.

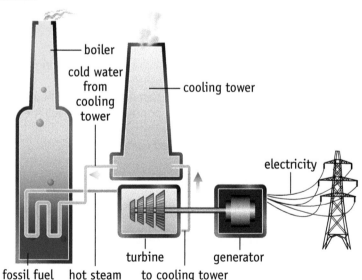

● Figure 2 Each stage in a fossil fuel power station shifts energy from one store to another

→ Power from natural gas

In a gas power station, things are a little different. The natural gas is burnt, and the exhaust gases drive the turbine directly. This generates electricity, but the turbine gets so hot it needs to be cooled down. The cooling water turns into steam and drives a second turbine, generating even more electricity (see Figure 3). This can make bring the efficiency of a gas power station to almost 60%, far higher than a coal or oil power station.

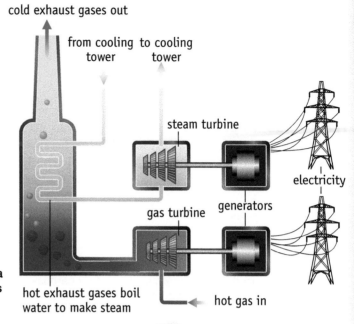

cold exhaust gases out

from cooling to cooling
tower   tower

steam turbine

electricity

generators

gas turbine

hot exhaust gases boil
water to make steam

hot gas in

● **Figure 3** A gas power station is sometimes called a **combined cycle power station** because it combines the output of two turbines

→ Why use fossil fuels?

Fossil fuels are used because they are very reliable. Unlike **wind** and **solar energy**, their availability does not depend on the weather. However, burning fossil fuels produces carbon dioxide as well as other forms of pollution. Increased carbon dioxide emissions are leading to climate change.

Fossil fuels are a non-renewable resource. Their current use is not sustainable. We need to look at alternative methods of electricity generation and more efficient systems.

→ Power from biomass

Some power stations burn wood from forests of fast-growing trees. These power stations work in the same way as ones that run on coal. Burning wood also puts carbon dioxide into the atmosphere, but the next generation of trees absorbs it again in photosynthesis, and so biomass is said to be **carbon neutral**. Other organic matter can be burnt to generate electricity – even poo!

Cars can run on **biofuels** – oil or alcohol made from crops grown on farms. As long as the same number of plants are grown each year, they will absorb the same amount of carbon dioxide as the previous year's fuel gave off.

→ Nuclear fuels

Nuclear fuels are not fossil fuels, but they will run out just the same. Nuclear power stations do not produce carbon dioxide. However, their **radioactive waste** is harmful and needs to be stored safely because it remains **radioactive** for thousands of years.

? Questions

1. How is the energy store of a fossil fuel released?
2. Which part of a power station produces electricity?
3. Nuclear power stations do not contribute to global warming, but many people oppose them. Give two reasons why.
4. **a)** Explain the difference between renewable and non-renewable resources.
   b) Give one example of each.
5. Try to find out the proportion of our country's energy needs that are currently met by using fossil fuels.

✎ Show you can...

Complete this task to show that you understand how electricity is generated in a fossil fuel power station.

Describe how the power station shown in Figure 2 works, making sure that you state what happens at each stage.

4.3 Energy from moving water

The destructive power of the sea is obvious after a hurricane or a tsunami, but some coastlines experience gradual erosion day after day, which eventually washes away whole streets of houses.

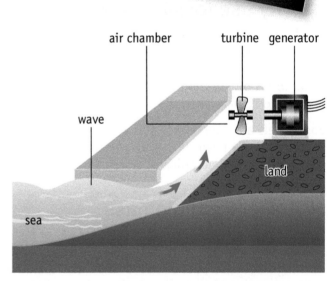

● **Figure 1** How do you think we could usefully harness the destructive energy of the sea?

Although the energy stores of the resources on these two pages are all different, they are linked by the fact that water is needed to release the energy in each case.

→ Wave power

Engineers are trying to find the best way to extract energy from sea waves. There are many different designs for wave power stations at present, and the technology needs to develop before this resource can be used effectively.

● **Figure 2** In an oscillating water column generator, waves compress air, which spins a turbine and generates electricity

→ Tidal power

A **tidal barrage** is a dam across a river mouth. Water passes through turbines in the dam as the tide comes in, and again as the tide goes out. This is a tried and tested method of generating electricity, but few barrages have been built because of worries about the ecological impact on the river estuary. Other designs, which look like underwater wind turbines, are being developed.

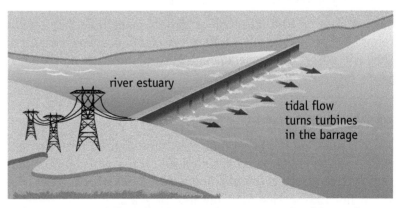

● **Figure 3** Tides follow the Moon, and are very predictable, so the power that they will generate is known well in advance

→ Hydroelectric power

Hydroelectric power (HEP) uses energy from the kinetic store of falling water to drive turbines. These power stations do not produce carbon dioxide and the resource will never run out – it is renewable. However, the large dams can flood habitats for people and wildlife.

→ Pumped storage

Pumped storage is not exactly an energy resource – it is a way of storing excess energy when people are using less than the power stations are producing. It looks like a hydroelectric power station that has been placed between high and low lakes, but has turbines that can be used as pumps. When there is 'spare' electricity, the turbines pump water up to the higher reservoir. When more electricity is needed, the water flows down through the turbines again, generating electricity.

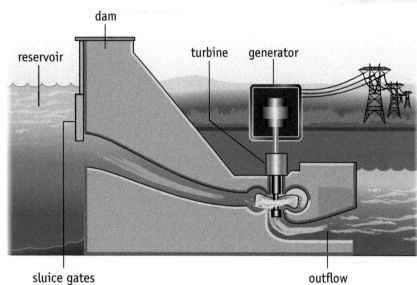

● **Figure 4** The **sluice gates** control the amount of water flowing through the turbine, and so the amount of electricity generated

? Questions

1 Explain why the UK is very well placed in Europe for wave energy.
2 All of the power stations on these two pages have two components in common. What are they?
3 At what time of day will a pumped storage system be:
  a) pumping water up to the top reservoir
  b) releasing water down to the bottom reservoir?
4 Which parts of the UK are more suited to hydroelectric power than others?
5 There are many designs for wave power stations, each using different methods of harnessing the kinetic store of the waves to make electricity. Find out about four different methods and make brief notes on how they work. Then choose one and describe it in detail. Include a labelled diagram.

✎ Show you can...

Complete this task to show that you understand the issues surrounding these ways of generating electricity.

List one argument for and one against using wave, tidal and hydroelectric power in the UK.

More renewable resources

Solar energy has been around for longer than humanity. Plants, from the smallest algae to the tallest tree, absorb sunlight to make sugars, which power their growth.

● **Figure 1** How could farmers have been the first people to own solar panels thousands of years ago?

→ Electricity from the Sun

Solar photovoltaic cells (PV cells) convert light directly into electricity. Energy radiated from the Sun is free and renewable. Solar cells are becoming cheaper and their efficiency is improving, but they still have a large **payback time** (the time it takes to 'repay' the original cost of the system). Also, some areas of the world receive less sunlight than others.

If a house has solar cells, it will still be connected to the **National Grid** (the network of pylons and cables that takes electricity around our country). The household buys electricity when it is using more than the solar cells are producing, but it sells electricity back to the Grid whenever the solar cells are producing more than the household uses.

→ Heat from the Sun

Solar panels are black to absorb as much of the Sun's heat radiation as possible. This heats up water that flows through the panels to provide free hot water, and even heating. In the summer the household boiler is not needed at all, and even in the winter the panels warm the water slightly, meaning the boiler needs to provide less energy.

● **Figure 2** A house can be fitted with solar panels and or photovoltaic cells, but the roof must face south (in the UK) to receive as much sunlight as possible

Power from the wind

Wind energy is renewable and does not produce carbon dioxide. It is not as expensive as solar power. **Wind farms** of many **wind turbines** take up a lot of space and they can be noisy up close. To solve these problems, some companies are now placing wind turbines offshore. Although it is not windy all of the time, there is always wind somewhere.

● **Figure 3** The larger a wind turbine is, the more efficient it is in capturing the wind's energy

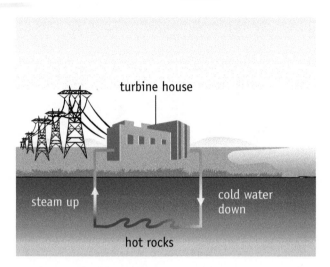

● **Figure 4** Geothermal power is more suited to Iceland or New Zealand than the UK, but one geothermal power station is now being built in Cornwall

Energy from underground

In **geothermal power** stations hot underground rocks turn water into steam to drive turbines and generate electricity. The heat rises from deep in the Earth's crust and is renewable, but there are only certain places on Earth that have a suitable geology.

? Questions

1 Some parts of the world are better suited to certain types of power generation than others. Do some research and find out three good locations for each of the following, explaining why they are well suited to that type of power generation:
   a) wind power
   b) geothermal power.

2 a) Explain the similarities and differences between solar photovoltaic cells and solar hot water panels.
   b) What are the advantages of having solar cells and solar panels on your roof?
   c) Are there any disadvantages of having solar cells and panels on your roof? List these too.

3 Some people say that biomass is a form of solar power. Explain why.

4 Do you think a household with solar PV cells would be taking electricity from the National Grid, or feeding electricity back into the grid, in the following situations:
   a) late in the evening, when the lights are on, the family are watching television, and the electric oven is cooking dinner
   b) during the day, when everyone is out at school or at work
   c) in the morning, when it is bright and sunny outside, and the family are boiling the kettle and using the toaster to make breakfast.

Show you can...

Complete this task to show that you understand what the energy sources are, for each of these types of power generation.

Write a sentence explaining why solar, wind and geothermal energy can be considered to be renewable.

Building scientific awareness

→ Following the Sun

In the northern hemisphere, solar panels and photovoltaic cells have to be placed so that they face south. This is because the Sun is always to the south of us in the sky. It rises in the east and sets in the west, and travels across the southern sky during the day. It does not travel directly overhead unless you live on the equator.

A solar cell will give its greatest output when the most light falls on it. You are going to investigate how the output of a solar cell changes as the angle to the light source changes.

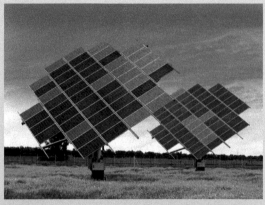

● **Figure 1** Why are these panels made to follow the Sun, rather than just stay stationary?

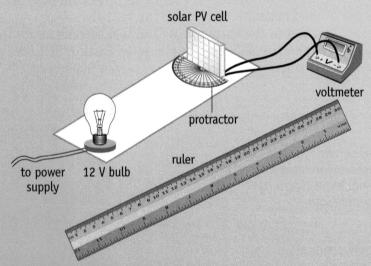

● **Figure 2** A sheet of paper helps when measuring the angle of the solar cell

Validity

To make your results **valid** you must show that the outcome is only dependent on one thing. You must write down which variable you are going to change, and which ones are going to be kept constant (it is a good idea to measure these during the experiment to prove that they are not changing).

Accuracy

To take **accurate** readings requires patience and care. Your eyes must be level with an instrument's scale when you are reading a value to ensure that you are reading the correct number, and all readings must be repeated to reduce the effect of **random errors**. Your final answer should be an average of your repeated readings.

Resolution

Resolution is all about how small a measurement to use. For example, if you were measuring your journey to school, it would be good enough to measure it to the nearest kilometre. The size of the lab, however, should be measured to the nearest metre, or even centimetre. And if you were to measure the length of a pencil, the nearest millimetre might be most appropriate.

Peer review

Scientists publish their results in scientific journals, which are read by their peers (other scientists in the field). This is called **peer review**. For your peer review, you can compare your results with those of your classmates to see whether they agree.

● **Figure 3** How can you ensure that your measurements are accurate?

Risk assessment

In every scientific activity you must always think ahead to evaluate any risks there might be. For example, what risks would there be in carrying out this experiment in a darkened room?

Your task

You are going to consider all of these issues and plan an investigation, which you will carry out if you have time. Your plan must:

- show how you will make sure your results are valid
- state how you will make each reading as accurate as possible
- state the resolution to which you will make your measurements
- show that you have considered any risks that the investigation might involve.

If you are able to carry out the investigation, your results must:

- show that you have only changed one **independent variable**, and state what the values of the **control variables** were
- show repeat readings and averages
- all be to the same, appropriate, resolution.

You must also discuss and compare your results with other groups in the class, to see if your findings back each other up.

You can write messages by placing magnetic letters on a fridge door. The door itself is not a magnet, but it is made from a magnetic material – one that is attracted to magnets. The two most common magnetic materials are iron and steel.

● **Figure 1** Could you stick magnetic letters to a fridge made of aluminium?

→ Magnetic materials

Most people will say that a magnet will pick up any metal. But if you ask them how to sort a steel can from an aluminium can for recycling, they will remember that the aluminium can will not stick to a magnet. The only **magnetic materials** are iron, steel, cobalt and nickel – and steel is only magnetic because it is mainly made of iron.

→ Attraction and repulsion

As well as attracting magnetic materials, magnets can exert forces on each other. These can be forces of both attraction and repulsion. The strongest regions of a magnet are called **poles**. One end of a bar magnet is a north pole and the other a south pole. A north pole and a south pole attract each other, but two north poles repel. Similarly two south poles repel each other.

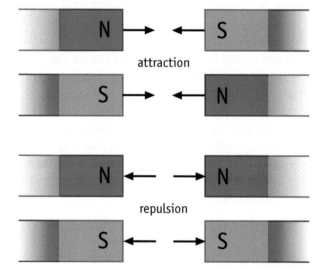

● **Figure 2** Unlike poles attract and like poles repel

→ Magnetising by stroking

The easiest way to make a magnet is simply to stroke a piece of steel, such as a nail, with another magnet. Make sure you always stroke it in the same direction with the same pole, and that you move the magnet away from the nail between strokes.

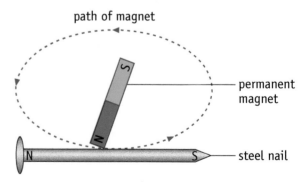

● **Figure 3** Stroking a steel nail with a magnetic pole causes it to become a magnet itself

→ Magnetic domains

If you break a magnet in half you do not get one piece of steel with a north pole and one with a south pole. You get two new, shorter, magnets.

You could carry on breaking the magnet into smaller pieces until all you had left were microscopic crystals of iron. Each crystal, called a **domain**, is a magnet in itself.

This also shows why there is a maximum strength to any magnet – when all of the domains are lined up the same way, the magnet cannot be made any stronger.

If the magnet is heated or dropped, the domains are shaken up, and can end up pointing in random directions again. This demagnetises the material.

● **Figure 4** The ends where the break happens form two new magnetic poles

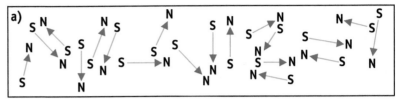

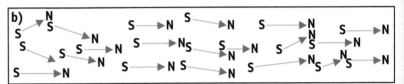

● **Figure 5 a)** If the domains in an iron bar are jumbled up, there is no overall magnetism to the bar. **b)** If enough domains are lined up the same way, the iron bar is magnetised

? Questions

1 What is the difference between a magnet and a magnetic material?

2 Can you make a magnet with just one pole? Explain.

3 Two pieces of iron attract each other. How can you tell whether they are both magnets?

4 **a)** What happens to the domains of an unmagnetised piece of steel when it is magnetised?

   b) What happens to the domains of a magnet when it loses its magnetism?

5 If I stroke a piece of steel with a magnet, it becomes magnetised. If I keep on stroking, or if I stroke it with an even stronger magnet, I find that I cannot make the steel's magnetism any stronger. Explain why.

✏ Show you can...

Complete this task to show that you understand the difference between magnets, magnetic materials and non-magnetic materials.

Copy the table below and separate these objects into the correct columns:

· a steel pen nib
· a brass belt buckle
· a pencil lead (made of graphite)
· an orienteering compass for navigating
· a mathematician's compass for drawing circles (made of iron)
· a plastic ruler
· a fridge magnet.

unaffected by a magnet	attracted to both poles of a magnet	repelled from one pole of a magnet
_____	_____	_____
_____	_____	_____
_____	_____	_____

5.2 Magnetic fields

Aurorae are formed when charged particles from the Sun hit the Earth's magnetic field. The field channels them to the north pole, where they make the atmosphere glow. Not only is the effect beautiful, but the process protects us from the worst of the Sun's radiation.

● **Figure 1** Where else on Earth can you see aurorae?

→ Magnetic fields

Magnets can attract paper clips without touching them. This is because there is a **magnetic field** surrounding the magnet. A magnetic field is a region in which the magnetic force acts. The Earth is like a giant magnet. Its magnetic field stretches over the whole of its surface and into space. Compasses are magnets, so they feel the force of the Earth's magnetic field, which makes them point due north. The 'north-seeking pole' of a compass ('north pole' for short) points towards the Earth's north pole.

→ Tracing a magnetic field

Magnetic fields have a particular shape. You can see this easily by using iron filings. Place a magnet under some paper and sprinkle the filings on the top. When you tap the paper, the filings fall back down in lines – these are called **field lines**.

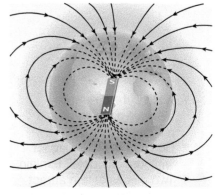

● **Figure 2** Unlike poles attract, so the magnetic pole found underneath the Earth's geographic north pole is actually a magnetic south pole

Field lines are drawn with arrows showing the direction that the magnet would push another magnetic north pole. You can see this happening by making a magnetised steel rod float in some water near to the magnet (see Figure 4).

As well as showing the direction of the magnetic force, the pattern of field lines also shows its strength. The magnetic force is strongest where the field lines are closest together.

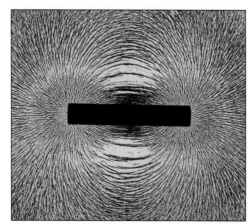

● **Figure 3** What shape would you see if you dipped the magnet into the iron filings? (Cover the magnet with clingfilm first!)

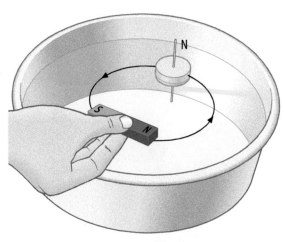

● **Figure 4** The rod moves in the direction of the field lines

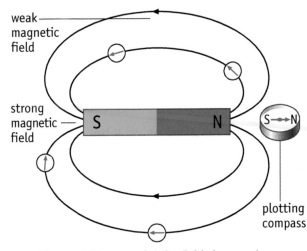

● **Figure 5** You can plot the field shape using a small compass

→ Magnetising by induction

If you pick up a steel paper clip with a magnet, you will find that you can make another paper clip stick to the first one. You can even make a whole chain of them, depending on the strength of the magnet. This is called **induced magnetism**.

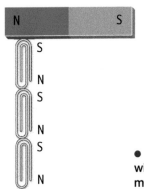

● **Figure 6** A strong magnetic field will line up the domains in a magnetic material, turning it into another magnet

→ Temporary and permanent magnets

Iron is described as a **soft magnetic material**, whereas steel is described as a **hard magnetic material**. When you magnetise steel it keeps its magnetism, but when you magnetise iron it loses its magnetism as soon as the magnetic field is removed.

? Questions

1 a) If the paper clips in Figure 6 were made of iron instead of steel, what would happen when the magnet was removed?
   b) Explain why this would happen.
2 How could you make a compass from a bar magnet and a piece of string?
3 If you were standing at the Earth's north pole, which way would your compass point?
4 This question is about how a magnet's field would change if it was made stronger.
   a) Sketch a diagram to show how Figure 5 would look if the magnet was replaced by a stronger one.
   b) What would happen to the rod and cork in Figure 4 if the magnet was replaced by a stronger one?

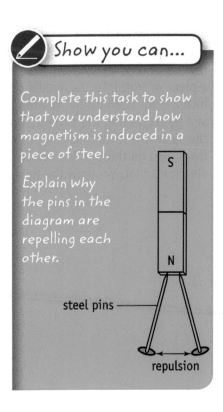

Show you can...

Complete this task to show that you understand how magnetism is induced in a piece of steel.

Explain why the pins in the diagram are repelling each other.

179

It would be no good using a bar magnet in a scrap metal yard. The magnet would be able to pick up the iron and steel but it would not be able to release it again. Instead we use electromagnets.

● **Figure 1** How do you think the scrap metal is released from the electromagnet?

→ Building an electromagnet

An **electromagnet** consists of a coil of lots of turns of wire wrapped around an iron **core**. When you pass an electric current through the coil, the electromagnet is magnetised.

Electromagnets lose their magnetism once the electric current is switched off. This means that you can pick up the scrap metal and put it down again where you want it to go.

→ Making it stronger

The magnetic field due to the current in a coil of wire (Figure 2) gets thousands of times stronger if you add a piece of iron into the middle of the coil. You can also increase its strength by winding more turns on the coil and by increasing the current. The magnetic field is the same shape as that of a bar magnet, but the **field lines** also go through the middle. To show a stronger electromagnet, we draw the field lines closer together.

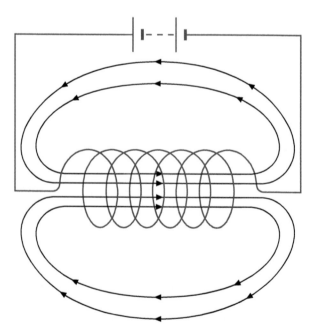

● **Figure 2** An electromagnet works because an electric current produces a magnetic field

You can investigate these factors by seeing how many iron nails an electromagnet can pick up. You could investigate the number of turns of wire and the size of the current as in Figure 3. Remember to change only one factor at a time and to control the other **variables**. How would you make your results as reliable as possible?

● **Figure 3** Change either the number of turns of wire or the current. Note down your answers and take repeat readings. You should only keep the electromagnet switched on for a few seconds at a time

→ Using electromagnets

Electromagnets have many uses. One of the most common is in an electric bell. Look at Figure 4. When the bell push is closed, current flows through the electromagnet. This magnetises, and pulls the hammer to hit the gong. It also pulls the contacts apart. This turns the electromagnet off, and so the hammer springs away from the gong, and closes the contact points again, allowing the current to flow once more. The hammer hits the gong again.

The whole process repeats until the bell push is released.

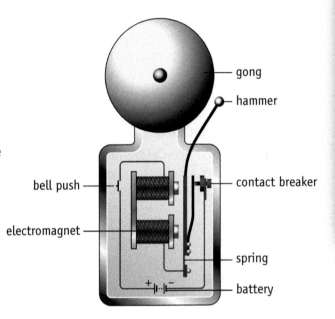

● **Figure 4** The electric bell relies on the electromagnet's ability to lose its magnetism, just as much as its ability to be magnetised

? Questions

1 A crane driver in a scrapyard is trying to pick up a car using an electromagnet to take it to the crusher, but he finds that the electromagnet is not strong enough.
   a) What can he do to make the electromagnet stronger?
   b) How does he release the car when it is above the crusher?
2 **a)** How would the electric bell function if the iron core of the electromagnet was replaced with a steel core?
   b) Explain why this happens.
3 An electromagnet in a scrapyard is often built with a U-shaped core. Why is this?
4 What simple change could you make to the electromagnet investigation to make the results more precise?
5 An electromagnet works because a wire carrying an electric current has a magnetic field. Do some research and find out the pattern of the magnetic field produced by:
   a) a single straight wire
   b) a flat coil
   c) a long coil that looks like a slinky spring (called a solenoid).

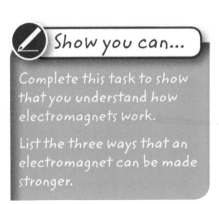

✎ *Show you can...*

Complete this task to show that you understand how electromagnets work.

List the three ways that an electromagnet can be made stronger.

181

5.4 More electromagnets

The Shanghai Maglev train uses electromagnets to lift the train above the track. The train can travel at 270 mph because friction is reduced. It takes under 8 minutes to travel the 18 mile journey to the city from the airport.

● **Figure 1** Which do you think is faster – the aeroplane that takes you to Shanghai airport, or the Maglev train from the airport to the city?

→ The loudspeaker

Loudspeakers depend upon **electromagnets** to work. Inside a loudspeaker an electromagnet sits in between the poles of a specially shaped permanent magnet. As the varying signal from the hi-fi passes through the electromagnet, it magnetises first one way, and then the other. This pushes it in and out of the permanent magnet, moving the loudspeaker cone with it, at the same frequency as the music.

● **Figure 2** Which part of the loudspeaker is the electromagnet?

Build your own loudspeaker

You can build a loudspeaker out of an A4 sheet of card and some insulated wire (see Figure 3).

1 Make a large cone and a tube and stick them together.
2 Wrap a metre or two of thin wire around it – this wire must be insulated to prevent a short circuit.
3 Connect the ends of the wire to a signal generator, and hang the cone from a clamp stand.
4 Finally, hold a bar magnet in another clamp stand and push it into the open end of the tube. When you switch the signal generator on you should hear a note.

You have made a loudspeaker.

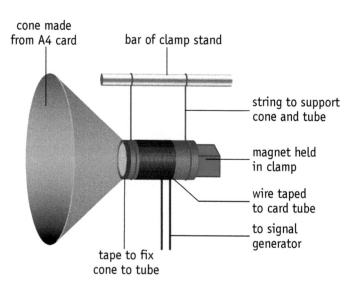

cone made from A4 card
bar of clamp stand
string to support cone and tube
magnet held in clamp
wire taped to card tube
to signal generator
tape to fix cone to tube

● **Figure 3** To make your loudspeaker louder you could use a stronger magnet, or wrap the wire around it more times

→ The electric motor

To make an electric motor spin, an electromagnet sits inside the field of a permanent magnet, as shown in Figure 4. The electromagnet turns on an axle when a current flows. The current in the electromagnet changes direction every half turn, so that the top of the electromagnet is always one pole, and the bottom is always the opposite pole. In this way the electromagnet is always pulled in the same direction, and spins continually.

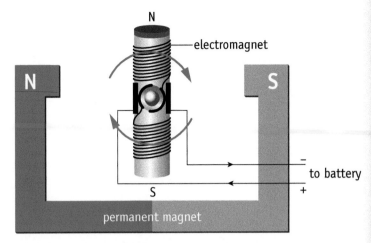

● **Figure 4** In this electric motor, the top of the electromagnet is always north, so it is always pulled to the right

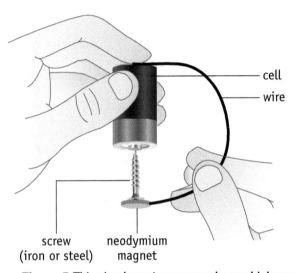

● **Figure 5** This simple motor can reach very high speeds

Build your own motor

You can build a motor (see Figure 5). Take a **neodymium magnet** and attach it to the thick end of an iron or steel screw (be careful – the magnetic field is very strong). Now hang the screw by magnetism from the bottom of a cell. Finally, hold a wire against the top of the cell, and gently touch the magnet with the other end. It should start to spin.

Wear goggles – the magnet spins very fast and the screw could fly off with no warning. Also, the wire can get hot. For this reason you should not use a rechargeable cell, as the current will get too high.

? Questions

1 Why must the coil in an electromagnet be wound using insulated wire?
2 The home-made loudspeaker in Figure 3 will not work if the permanent magnet is not present. Explain why.
3 The direction that the electric motor spins will change if you change the direction of the electric current (swap + and – at the power supply). What else could you do to change the direction in which the motor spins?
4 a) You cannot take a real loudspeaker apart to change the magnet or the coil every time you want to turn the volume up or down. What happens in a real loudspeaker to make it:
   i) quieter   ii) louder?
  b) What three things could you do to the motor in Figure 4 to make it:
   i) spin faster   ii) spin slower?

✎ Show you can...

Complete this task to show that you understand the importance of electromagnets in the modern world.

Make a list of as many things as you can that use electromagnets, and of those that use permanent magnets. Which is the bigger list?

Calculating scientifically

→ **Pole reversal**

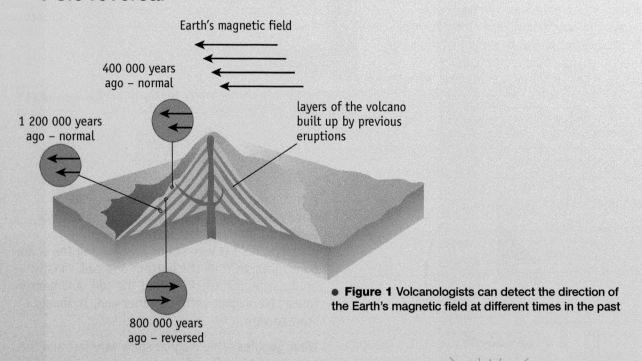

Earth's magnetic field

400 000 years
ago – normal

1 200 000 years
ago – normal

layers of the volcano
built up by previous
eruptions

800 000 years
ago – reversed

● **Figure 1** Volcanologists can detect the direction of
the Earth's magnetic field at different times in the past

A volcano is made of layers of **basalt** rock,
which is solidified **lava**. Each time the volcano
erupts, it lays down another layer of basalt.

Volcanologists (scientists who study
volcanoes) have discovered that when lava
becomes solid rock, any magnetic particles in
the lava freeze in the direction of the Earth's
magnetic field.

When they measured the magnetic field of
basalt they found that the magnetic field of
older basalt pointed in the opposite direction
to basalt that had formed more recently. The
only way that this could have happened is if
the Earth's magnetic field had flipped in the
past. When they investigated further, they
found that this had happened more than once.

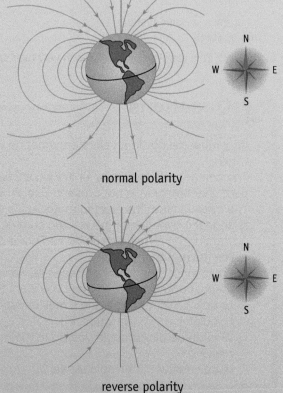

normal polarity

reverse polarity

● **Figure 2** The Earth's magnetic
field has flipped many times in
the past

Timescales

To work out how often the Earth's magnetic field has flipped, and if it is likely to happen again soon, the volcanologists had to find out the following:

t – the average time between eruptions

n – the number of eruptions that will make a layer of basalt 1 km deep

d – the depth of basalt between magnetic field reversals (measured in km).

Combining these three in the correct way enabled them to find out the average time between magnetic field reversals.

Your task

1 If t is the time between eruptions and n is the number of eruptions that makes a 1 km thick layer of basalt, which of the followng is the average time to form this layer of basalt?
   - $n \times t$
   - $\dfrac{n}{t}$
   - $n + t$
   - $n - t$

2 We can call this average time to form a 1 km layer of basalt k. If the depth of basalt between field reversals is d, then which of these is the average time between field reversals?
   - $k \times d$
   - $\dfrac{k}{d}$
   - $k + d$
   - $k - d$

3 The volcanologists consulted the history books, and found that the average time between eruptions was 100 years. They measured the depth of the most recent eruptions, and found that 2000 eruptions were needed to produce 1 km depth of basalt. They drilled into the volcano and measured the magnetic field of the rock at different depths. This told them that the depth of basalt between magnetic field reversals was 2 km on average. Use your equations, and the values given above, to find out the average time between the reversals in the Earth's magnetic field. Here are the figures again:
   $t = 100$ years
   $n = 2000$ eruptions per km
   $d = 2$ km

A force is needed to make a moving object change direction. In the Wall of Death, the motorbike would travel in a straight line if it were not for the wall pushing against it.

● **Figure 1** In which direction is the force that makes the motorcyclists move in a circle?

→ Changing speed

The cars in Figure 2 are moving in a straight line. To make the red car **accelerate** (speed up) the forces on it must be unbalanced in the direction of its motion. The force from the engine creates an unbalanced force. The car starts to move and gets faster and faster.

● **Figure 2** Unbalanced forces are needed to speed up or slow down

Friction forces produced by a car's moving parts and air resistance oppose the car's motion when it is moving. The driving force on the yellow car in one direction matches the opposing forces in the other direction, so the forces are balanced. The car travels at a steady speed.

To make the blue car **decelerate** (slow down) the forces must be unbalanced in the opposite direction to its motion. The force of the brakes (plus air resistance) provides this unbalanced force.

→ Changing direction

An unbalanced force can also make a moving object change direction. The Earth's gravity is always pulling on the Moon, but the Moon does not get any faster. Gravity actually keeps the Moon in its orbit.

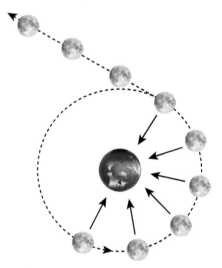

● **Figure 3** The Moon would carry on in a straight line if it were not for the Earth's gravity constantly changing its direction

→ Investigating how force affects speed

Look at Figure 4. When the ball is placed on the runway, its weight acts down the slope, while friction acts up the slope. If the slope is steep enough, then there is an overall force down the slope. This makes the ball start to move and then speed up. You can get an idea of how much the ball speeds up by timing how long it takes to roll down the slope.

When you carry out an investigation you should predict what would happen if you changed some of the variables. Making **predictions** based on a theory and then testing them by experiment is an important process in science. If the results agree with your prediction then it suggests that your theory is correct. How would the results change if you made the slope twice as steep?

Once the ball reaches the horizontal bench it begins to slow down. This is because the force of friction acts in the opposite direction to the ball's motion. If there were no friction at all, what would happen?

● **Figure 4** You can investigate how forces affect speed by rolling a ball down a slope

● **Figure 5** These forces are applied to a car that is initially stationary

● **Figure 6** Forces on a motorbike that is travelling forwards

? Questions

1 Look at Figure 5. What will happen to each of the cars when the forces shown are applied?
2 What force makes the following objects change direction instead of travelling in a straight line?
   a) a ball bouncing off a wall
   b) a yo-yo reaching the end of its string
   c) the Earth orbiting the Sun
3 Look at Figure 6. What will happen to each of the motorbikes when the forces shown are applied?

✎ Show you can...

Complete this task to show that you understand how an unbalanced force can affect the speed of an object.

Copy and complete the following sentences:

1 An unbalanced force acting on a stationary object will cause it to ....
2 An unbalanced force acting on an object in the same direction as it is moving will make it ....
3 An unbalanced force acting on an object in the opposite direction to its motion will make it ....
4 An unbalanced force acting on an object at 90° to the object's motion will make it ....

A speed camera takes two photos of a speeding car. The marks on the road measure the distance that the car travels between the photos. Dividing this distance by the time between the photos gives the average speed.

● **Figure 1** If the car goes faster, will the distance it travels between the two photos be larger or smaller?

→ Calculating speed

To work out how fast something travels over a certain distance, you need to use the equation:

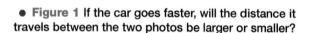

$$\text{average speed} = \frac{\text{distance travelled}}{\text{time taken}}$$

For example, if a car travels 25 m during 0.5 s then:

$$\text{average speed} = \frac{25\,\text{m}}{0.5\,\text{s}} = 50\,\text{m/s}$$
$$= 180\,\text{km/h} \ (112\,\text{mph})$$

The units for **speed** come from the units for distance and time. If the distance is in metres (m) and the time is in seconds (s), then the speed is in metres per second (m/s).

→ Measuring speed

The light gate measures how long it takes for a piece of card to pass through it. It does this by timing how long the light beam was blocked by the card. The distance the trolley travels in that time is equal to the length of the card.

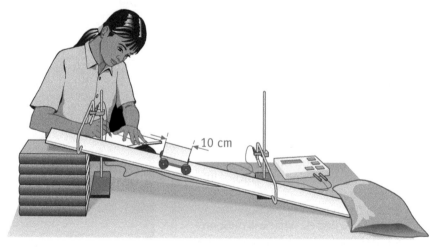

● **Figure 2** You can measure the average speed of a trolley in a laboratory using **light gates**

→ Relative speed

Imagine you are on a train travelling at 100 mph, watching a car through the carriage window (see Figure 3). Because the train is travelling faster than the car, it will be moving backwards relative to you, even though it may be driving at 70 mph.

● **Figure 3** Your relative speed can be higher or lower than your actual speed

→ Finding speed from a graph

On a graph of distance against time, the distance an object has travelled is plotted on the vertical axis, and the time it has taken to travel this distance is plotted along the horizontal axis.

If we divide the amount that the line on the graph has risen by the amount it has gone along we get the **gradient** of the graph. This is the same as the speed of the object.

Figure 4 is a distance–time graph for a car that was speeding until the driver noticed a speed camera up ahead and slowed down. The gradient starts large, indicating a fast speed. As the gradient decreases, so does the car's speed.

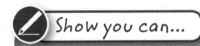

● **Figure 4** A graph of distance against time can also be used to find an object's speed

? Questions

1 Copy the following paragraph using the correct words from the brackets:
   Two sprinters ran a 100 metre race. The faster runner covered a (longer/shorter) distance in the same time that the slower runner covered a (longer/shorter) distance. When they found their final times for running 100 m, the faster runner had taken a (longer/shorter) time, while the slower runner had taken a (longer/shorter) time.

2 In the experiment described on page 188 the card attached to the trolley was 10 cm long. How fast was the trolley travelling, in cm/s, if the card blocked the light beam for:
   a) 2 seconds
   b) 0.5 seconds?

3 What would the distance–time graph in Figure 4 look like if the car started travelling slowly, but then sped up?

4 The Marathon of the Sands is the equivalent of six marathons, run across the Sahara desert.
   a) Write down the equation a runner would use to work out their average speed, if they knew the distance they had travelled and the time they had taken.
   b) How would they rearrange this equation to calculate the distance they had travelled if they knew their speed and the time they had been running for?

✎ Show you can...

Complete this task to show that you understand how to find out the speed of an object.

Write a description of how you would find the speeds of your friends walking, jogging and sprinting, listing any equipment you would use, and describing fully how you would use it.

In the London 2012 Olympics divers plunged from a platform 10 m high, hitting the water at speeds of up to 30 mph (14 m/s).

● **Figure 1** Would divers of different masses fall at different speeds?

→ Mass and weight

There is an important difference between mass and weight.

- The **mass** of an object tells us how much matter there is. Mass is measured in kilograms (kg).
- The **weight** of an object is the force of gravity acting on it. Weight is measured in **newtons** (N) because it is a **force**.

Every kilogram of matter on the Earth is pulled down by gravity with a force of 10 newtons. Because of this, your weight in newtons is 10 times your mass in kilograms.

Weight (W) is equal to mass (m) multiplied by the Earth's **gravitational field strength** (g):

$$W = m \times g$$

The gravitational field strength is 10 N/kg here on Earth. On the Moon it is less, because the Moon is smaller than the Earth.

Imagine an astronaut with a mass of 72 kg. On Earth his weight would be 720 N. If he goes to the Moon, his mass will still be 72 kg (as he has not lost any matter), but his weight will only be 120 N. The Moon exerts one-sixth of the gravitational force of the Earth.

● **Figure 2** Would you weigh more or less on the Moon?

→ Free-fall

An object with more mass is pulled more strongly by gravity, so you would think that it would accelerate more than a light object. However, you need to push/pull harder on the more massive object than you would on the light one in order to accelerate them both by the same amount. These effects cancel out, and so the acceleration due to gravity does not change, whatever the mass of the object.

→ Using a speed–time graph

To make a graph of speed against time, the speed of the object is plotted on the vertical axis, while time is plotted along the horizontal axis. The gradient of the speed–time graph tells us the object's acceleration; the steeper the gradient, the greater the acceleration.

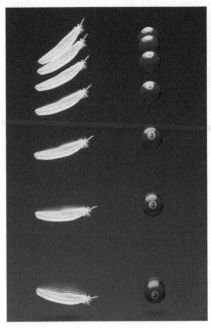

● **Figure 3** In a **vacuum** a feather falls to the ground with the same acceleration as a pool ball

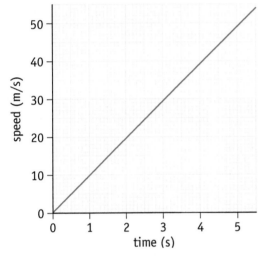

● **Figure 4** A graph of speed against time can be used to find an object's acceleration

Figure 4 is a speed–time graph for the feather and the pool ball in Figure 3. The gradient is constant, showing that (as long as air resistance is zero) their acceleration is also constant.

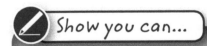

Questions

1 If you were to drop a pool ball and a feather in the classroom, the feather would hit the ground last. Explain why.
2 If your mass is 60 kg, what is your weight:
   a) on the Earth
   b) on the Moon?
3 In 1971, while standing on the Moon's surface, Apollo astronaut David Scott simultaneously dropped a hammer and a feather. They fell at the same rate, and reached the ground at the same time. Which of the following facts are due to the Moon's gravity being weaker than the Earth's, and which are due to the fact that the Moon has no air?
   a) The hammer and the feather hit the ground at the same time.
   b) The hammer fell more slowly than it would have done on Earth.
   c) The feather fell faster than it would have done on Earth.

Show you can...

Complete this task to show that you understand how objects accelerate due to gravity.

Explain how the photograph of the feather and ball would differ if they were dropped on the Moon.

6.4 Streamlining

In the **Mud Pit Belly Flop** contest, contestants try to make the biggest splash possible by presenting a large surface area to the mud.

● **Figure 1** How does this woman's body position differ from the high divers on page 190?

→ Streamlining in nature

The kingfisher's body is shaped to make the frictional force (resistance) of air and water very small. This makes its speed in the water as great as possible. **Streamlining** reduces friction forces (**drag**) by creating a smooth surface and an aerodynamic shape. The dandelion seeds are shaped to maximise **air resistance**, so they are carried far by the wind.

● **Figure 2** Why is the kingfisher's body streamlined?

● **Figure 3** Why are dandelion seeds not streamlined?

→ The effect of speed

As you try to move faster through the air, its resistance increases. It is easy to pass through the air at walking pace, but you begin to feel the effect of air resistance if you cycle very quickly.

At 150 mph, this speed-skier's posture and clothes need to be very streamlined. The air flows very smoothly around her. Without streamlining the air would not flow smoothly. This is called **turbulent flow**. Turbulent flow makes the force of air resistance much bigger. Air resistance acts in the opposite direction to motion, so the skier would not be able to ski as fast.

● **Figure 4** Would the skier go as fast if she were standing up?

→ Increasing air resistance

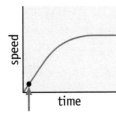

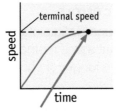

At first, this skydiver is not falling very fast. The air resistance is small. The force of gravity (the skydiver's weight) provides an unbalanced force. Her acceleration is shown by the gradient of the graph.

The skydiver is now falling faster. Air resistance has got larger. There is still an unbalanced force downwards, but not as large. She continues to accelerate, but not as quickly, so the gradient is not as big.

The skydiver is falling so fast that her air resistance is as large as the force of gravity. The forces are balanced and she falls at constant speed, called her **terminal speed**. The gradient is zero because her acceleration is zero.

● **Figure 5** The faster she falls, the less the skydiver accelerates

? Questions

1 a) Make two lists, one headed 'Very streamlined' and the other headed 'Not streamlined'. Sort the following animals into these two categories, and write them under the correct heading in your book.

   bear   dolphin   snail   shark   koala
   wolf   antelope   jellyfish   elephant   cheetah

   b) Make two more lists, one headed 'Fast' and the other headed 'Slow'. Group the animals in question 1 under these two headings in your book.

   c) Now explain any similarities and differences between the lists.

2 The following objects were dropped from an aeroplane. Some reached a higher terminal speed than others. List them in order of their terminal speeds, fastest first.

   dandelion seed   marble   china bowl
   handkerchief   suitcase

3 The skydiver in Figure 5 is holding her body in a shape that maximises her air resistance.

   a) How could she change her body shape to increase her acceleration?

   b) Explain why she slows down when her parachute opens, in terms of streamlining and air resistance.

Complete this task to show that you understand how and why an object is streamlined.

Explain the difference in the shapes of a toy rocket and a kite.

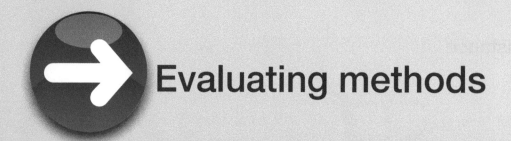

Evaluating methods

→ Dive right in!

Ethan and Sofia wanted to find out the time it took high divers to fall 10 m. They sat at the poolside with their stopwatches, and noted down the time it took for each diver to hit the water after leaving the platform.

Looking for errors

When Ethan and Sofia plotted the times on a graph, they found that some methods of measurement generally gave longer or shorter times than others.

Random errors may be caused by a faulty technique in taking measurements, or by the differences in human reaction times. They can cause readings to be either too high or too low, and by varying amounts. Random errors can be detected by taking a large number of readings. Averaging these readings compensates for random errors.

When measuring the volume of a liquid:

- viewing from above will make the reading appear greater than it really is.
- viewing from below will make the reading appear lower than it really is.
- viewing at the level of the liquid will give the correct reading.

Systematic errors cause all of the readings taken to be shifted above or below the **true value**. Taking an average of these readings will not help, because the average will still be higher or lower than the true value. A systematic error can occur when using a wrongly **calibrated** instrument, or when persistently making a measurement in the wrong way.

A **zero error** is a type of systematic error. It is caused by using instruments that have a false zero, e.g. if a top pan balance shows a reading when there is nothing placed on the pan.

● **Figure 1** How do you think a more streamlined dive would affect the time it took to hit the water?

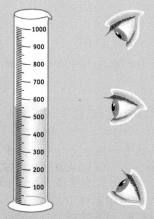

● **Figure 2** Reading the volume of a liquid in a measuring cylinder can involve random errors

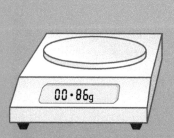

● **Figure 3** The 20 g block appears to weigh 0.86 g more than its true mass

Ethan and Sofia discussed their results, and the possible reasons for the spread of points on their graphs.

- Some measurements were for dives in which the divers started crouched down.
- Other measurements were for dives in which the divers started standing up.
- Some were for dives with the divers in more streamlined positions.
- Others were for dives with the divers in less streamlined positions.
- One set of measurements was taken by stopping the stopwatches when they heard the splash of the diver hitting the water, instead of when they saw the diver enter the water.

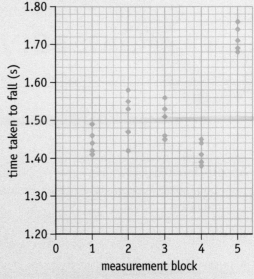

● **Figure 4** A graph showing the times for high divers to fall 10 m. Points in each measurement block are times measured in the same way

Task 1

Discuss with your partner which of these measurement methods could have given rise to each block of results in the graph.

> 1 List in your book the factors that could have made some time measurements longer or shorter than others, and write down, with reasons, which block is explained by which of the possible reasons above.
> 2 Also write down whether you think each is a random or a systematic error.

Task 2

Now think about measurements you make in the lab.

> 1 Make a list of the things that you are often required to measure, the instrument you use to make the measurement, and how a random or a systematic error could creep in.
> 2 Also note down how you could guard against these errors occurring.
> The following ideas are given to start you off:
> - Measuring my height using a metre ruler...
> - Measuring the temperature of water in a beaker using a thermometer...

The errors in Task 1 were all systematic errors. What could have been a source of random errors in these timings?

● **Figure 5** If you stopped the stopwatch when you heard the splash, would your time be correct?

Task 3

> 1 Now choose the measurement block from the graph that you think gives the best set of results.
> 2 Read off the values from the graph, and find the average of these to give a final time that accounts for random errors.

Glossary

Absorption In digestion, the movement of small food molecules from the small intestine into the blood

Accelerate To speed up

Accuracy An accurate result is one that is close to the true value

Acid A substance with a pH less than 7

Acid rain Rain containing dissolved sulfur dioxide, from coal power station emissions, making it acidic, with harmful effects on the environment

Adaptation A feature that helps an organism to survive in its habitat

Air resistance A contact force between the air and an object moving through it, which opposes the object's motion

Alga (plural algae) A simple plant

Alkali A substance with a pH greater than 7

Ammeter A device used to measure electrical current

Amnion The protective sac that surrounds a developing foetus

Amniotic fluid Fluid around the foetus in the amnion that supports it and protects it from bumps

Amoeba A unicellular organism that lives in ponds

Ampere (A) The unit for current ('amp' for short)

Anaemia Disease caused by shortage of iron in the diet

Anther The part of a flower's stamen that contains pollen grains

Applied force A contact force between two objects, such as a push or a pull

Atom A small particle that makes up all substances

Balanced diet A diet containing the seven food types in the right proportions

Balanced forces Two forces of the same size acting on the same object in opposite directions, giving a zero resultant force

Basalt Solid rock formed from cooled lava

Battery Two or more cells put together to produce a higher potential difference (p.d.)

Benedict's solution Solution used to test for glucose; if glucose is present it turns green, orange or red when heated

Bioaccumulation The increase in concentration of a chemical in organisms along a food chain

Biodegradable plastic Plastic made from plants that will eventually be broken down by micro-organisms

Biofuel A fuel, such as oil or alcohol, which is made from a crop grown on a farm; biofuels can be used to replace the fuels we get from crude oil

Biomass In Biology: the mass of living material

Biomass In Physics: A fuel that is composed of (recently) living organisms, such as wood

Biuret reagent A solution used to test for proteins; if protein is present it turns lilac or purple

Bladder The body organ that stores urine

Boiling The change of state when a liquid rapidly becomes a gas

Calibrated When a piece of measuring equipment is checked against a known standard value to ensure its accuracy

Carbohydrate A compound obtained from food that is needed as a short-term energy store

Carbon neutral Any energy resource that takes in as much carbon dioxide from the atmosphere when it is formed as it gives out when it is used

Carnivore An animal that eats other animals

Carpel The female part of a flower. It has three sections: the ovary, the style and the stigma

Catalyst A compound that speeds up a chemical reaction but is unchanged itself

Cell In Biology: the building block of living organisms

Cell In Physics: a simple electrical power supply that uses a chemical reaction to push charge around a circuit

Cell membrane Controls what enters and leaves a cell in an organism

Cell wall The outer part of a plant cell that helps to support the cell

Cervix The entrance to the womb (uterus)

Chemical reaction A change in which one or more new chemical substances are made

Chemical store Energy that can be released by a chemical reaction

Chemosynthesis A process used by some bacteria to obtain energy from simple chemical compounds

Chlorophyll A green substance that plants use to absorb light energy

Chloroplast The structure in a plant cell in which photosynthesis takes place

Chromatography A technique used to separate mixtures of coloured compounds

Circuit components Items such as cells, switches, bulbs and resistors that can be used to construct an electrical circuit

Classification Sorting organisms into groups based on their similarities and differences

Climate change Changes in the pattern of weather that a region usually experiences; seen around the world as a result of global warming

Coal A rock formed from ancient trees and other plants that were buried under layers of sediment

Combined cycle power station A power station that burns gas to drive one turbine and uses steam, produced by its own cooling water, to drive another

Compact fluorescent bulb An energy-efficient light bulb that works by passing an electric current through a low pressure gas in a glass tube

Competition When resources are limited, organisms will compete with each other for them; those that are better adapted are more likely to survive

Compound A substance made from two or more different types of particles that are chemically combined

Compressibility A property of a substance that means it can take up a smaller volume when pressure is placed on it

Concentrated Containing high amounts of a substance

Concentration A measure of the amount of solute (solid) dissolved in a solvent (liquid)

Condensation The change of state when a gas becomes a liquid

Conduction The transfer of heat by particle vibrations passing along a material; or the passing of a current through an electrical conductor

Conductor (electrical) A material that allows electric current to pass through it – it has low electrical resistance

Conductor (thermal) A material that allows heat to be transferred easily through it

Consumer An animal that eats another organism, either a plant or an animal

Continuous variable A measurement taken in an experiment that can have any value (see Variable)

Continuous variation A type of variation in which the data do not fit into groups, so they can have any value

Control variables The things that must be kept the same in an experiment to make sure they do not affect the dependent variable

Convection The transfer of heat when warm air/water rises and cooler air/water sinks, due to differences in density

Cooperation When organisms work together to survive, for example to obtain food

Core The material used in the centre of an electromagnet; usually iron

Corrosive Can destroy living tissue

Cross-pollination The transfer of pollen (for example by an insect) from one flower to another flower

Crude oil A fossil fuel formed from the decay of dead sea creatures that fell to the sea floor and were covered by layers of sediment

Cytoplasm The part of an organism's cell where many chemical reactions happen

DDT A pesticide that takes a long time to break down and has now been banned in many countries

Decelerate To slow down

Deficiency disease A disease caused by lack of a mineral or vitamin in the diet

Deforestation Cutting down large areas of forest for agriculture, timber, roads and housing

Density The ratio of an object's mass to its volume; the greater the density, the more 'compact' an object is

Dependent variable The outcome of an experiment – the variable that changes, depending on the other variables in the experiment

Diatomic An element that is made up of two atoms combined as a molecule

Diffusion The movement of particles from an area of high concentration to an area of lower concentration

Digestion The breakdown of large molecules into small ones so that they can be absorbed into the blood

Digestive system The group of body organs that digest and absorb food

Dilute Contains a low proportion of particles of a substance, as water has been added

Discontinuous variation A type of variation that fits into groups; there are no in-between values

Dissipated When energy becomes more spread out and less useful for doing work, for example when energy transfers from the thermal store of a house to the thermal store of the environment

Dissolves The process in which solid particles break apart and mix with liquid particles

Domain The microscopic crystalline structure of a magnetic material; if the domains are aligned in the same direction, the material will be magnetised; if they are aligned in different directions, the material will not be magnetised

Drag Air resistance or water resistance

Ecologist Scientist who studies different environments and the organisms that live in them

Ecosystem A set of living organisms and non-living factors (such as temperature, light, water and shelter) that interact with each other

Effervescence The escape of gas from a solution, recognised by bubbles rising

Efficiency The amount of energy usefully given out by an appliance as a proportion of the total amount of energy it takes in

Egg The gamete (sex cell) produced by a woman; also called an ovum (plural ova)

Ejaculation The release of semen, the liquid containing sperm, from a man's penis into a woman's vagina

Elastic A property of a substance that means it will return to its original shape after being stretched

Elastic deformation The temporary stretching or squashing of an object by a force acting on it; the amount of deformation is proportional to the size of the force, and it returns to zero when the force is removed

Elastic store The energy stored by stretching or bending an object

Electric current (I) The rate of flow of electric charge

Electrical store The energy stored by the attraction or repulsion of electrical charges

Electrical work Shifting energy from one store to another, by use of an electrical current

Electrolysis A process of splitting compounds by using electricity

Electromagnet A device made of a coil of wire wrapped around a soft iron core, which can be magnetised by making a current flow through the wire

Electromagnetic radiation Waves of varying wavelengths that can transfer energy through a vacuum, from the longest wavelength radio waves to the shortest wavelength gamma rays, and including infrared, visible light and ultraviolet radiation

Electron microscope A very powerful microscope with a higher magnification than a light microscope

Electrostatic force A non-contact force between two electrically charged objects that can attract or repel

Element A substance made from just one type of atoms

Embryo A ball of cells formed from a fertilised egg

Endoscope An instrument used by doctors to see inside the body

Energy resource Something that can be harnessed to provide energy (mainly, but not only, referring to electricity generation)

Energy store The way an object holds its energy

Environmental variation Variation in organisms caused by environmental factors such as diet

Enzyme A biological catalyst that speeds up chemical reactions in the body

Equilibrium An object in equilibrium has no unbalanced forces acting on it

Erosion When the wind, rivers or the sea wear away rocks

Euglena A unicellular organism that lives in ponds

Evaporation The change of state when liquid particles change into a gas; this happens slowly and occurs below the boiling point

Expansion When a substance gets bigger in size without any change in mass

Extension How much a spring or elastic band changes in length when stretched

External fertilisation The fusion of egg and sperm nuclei outside an organism's body

Extinct When the last member of a species has died

Fat A compound used as a long-term store of energy in the body and that forms an essential part of cell membranes and nerve cells

Fertilisation The fusion of the nucleus of a female sex cell (gamete) with the nucleus of a male sex cell

Fertiliser A substance used by gardeners and farmers that helps plants to grow

Fertility drug A drug that stimulates the release of eggs in women

Fibre Cellulose from the cell walls of plants, which helps the movement of food through the human digestive system

Field lines Lines that show the direction of a magnetic field; the closeness of the lines also shows the strength of the field

Filament In a flower, the part of the stamen that holds up the anther

Filament bulb An old-fashioned type of light bulb, which passes current through a thin metal wire to heat it up and make it glow brightly (very wasteful of energy)

Flagellum The whip-like structure that *Euglena* uses to move through water

Foetus The developing baby inside the womb after 8 weeks of pregnancy

Food chain A diagram that shows the flow of energy and nutrients through the organisms in a habitat

Food web A diagram that shows the links between different food chains in a habitat

Force An action that can stretch or compress an object, or cause it to speed up, slow down or change its direction of motion

Fossil fuels Fuels formed from organic matter over hundreds of millions of years: coal, crude oil and natural gas

Freezing The change of state when a liquid becomes a solid

Friction A force between two touching surfaces that are moving relative to one another. It can make them grip each other usefully, or it can be a nuisance, slowing motion down

Fuel A substance that can be reacted with oxygen, through burning or respiration, to release energy from its chemical store

Gametes Body cells involved in reproduction: in humans, sperm cells and egg cells

Gas pressure The force exerted when gas particles collide with the walls of a container

Gene A section of DNA that controls one inherited characteristic of an organism

Generator The next stage in a power station after the turbines; inside the generator coils of wire spin through a magnetic field to make electricity

Genetic engineering The transfer of genes from one organism to another in order to transfer a useful characteristic

Genetic variation Variation in organisms that is controlled by genes; also called 'inherited variation'

Geothermal power Using the heat from rocks deep underground to generate electricity

Germinate When a seed starts to develop into a plant by growing a root and a shoot

Gestation The length of time from the fertilisation of an egg to the birth of the baby; about 40 weeks in humans

Global warming The warming of the Earth due to the increase in greenhouse gases

Gradient How steep the slope of a line graph is

Gravitational field strength The force that the Earth exerts on each kilogram of matter on its surface

Gravitational store Energy stored due to an object's height, which can be released by letting the object fall

Gravity The force of attraction between the Earth and another object (see also Weight)

Guard cell One of a pair of plant cells that control the size of a stoma in a leaf

Gut flora Bacteria that live in the digestive system

Habitat The place where an organism lives

Haemophilia Inherited disease that increases the time it takes for blood to clot

Hard magnetic material A material such as steel that can be used to make a permanent magnet

Heat transfer The transfer of energy between the thermal stores of two objects

Herbicides Chemicals used to kill weeds

Herbivore An animal that eats plants

Histologist A scientist who studies cells

Hooke's law Describes how an object, such as a spring, extends or compresses when a force is applied to it

Hormones Chemicals produced by glands that affect cells in a different part of the body

Hydroelectric power (HEP) Using running water to generate electricity; usually involves a dam built to store a large reservoir of water, which is released through turbines when electricity is needed

Hydroponics The method of growing plants in a solution of mineral salts rather than in soil

Hydrothermal vent A hot-water spring found in deep oceans

Immiscible Liquids that will not mix with each other

Implantation The sinking of an embryo into the thickened lining of the womb (uterus)

Impurity A substance that is different from the intended material

Incomplete combustion When a fuel burns without enough oxygen to react completely; the products are carbon monoxide (CO) and water

Independent variable The input to an experiment – the variable that you change, to find out what happens

Indicator A chemical substance that changes colour in an acid or an alkali

Induced magnetism Magnetising an object by placing it in the field of another magnet

Infertility When a couple cannot have a baby

Infrared radiation Heat transfer by an electromagnetic wave with a wavelength just outside the visible spectrum, beyond red

Inherit To receive a characteristic from parents due to genes that are passed from parents to offspring

Inherited variation Variation in organisms that is controlled by genes; also called genetic variation

Insoluble A solid that does not dissolve

Insulator (electrical) A material that does not conduct electricity – it has a very high resistance; always a non-metal

Internal fertilisation The fusion of egg and sperm nuclei inside the female's body

Invertebrate An animal without a backbone

Investigation A practical activity designed to test a prediction by finding the effect on one variable when another variable is changed

Iodine solution Solution used to test for starch; if starch is present it turns black

Irritant Something that causes soreness or itching when in contact with living tissue

IVF (in-vitro fertilisation) Fertility treatment that involves fertilising an egg outside the woman's body, then implanting the embryo formed into her womb (uterus)

Kilocalorie A measure of energy, commonly used in food labelling and dietary advice, which is equal to 4184 joules

Kilojoule One thousand joules

Kilowatt One thousand watts, or one thousand joules per second

Kinetic store The energy of an object due to its motion

Large intestine The part of the digestive system that absorbs water and forms faeces from undigested food

Lava Liquid rock from a volcano

Law of conservation of energy The principle that energy can neither be created nor destroyed, it is just transferred from one store to another

Light Radiation that can be detected by the human eye

Light-emitting diode (LED) The most energy-efficient type of light bulb

Light gate Laboratory equipment for measuring motion, which uses a light beam to measure the time for an object to pass across the light beam

Litmus A natural substance used as an indicator; acids turn litmus red, and alkalis turn litmus blue

Load A downwards force on an object due to the weight of the items it supports

Lubricant A liquid that is placed between two surfaces to reduce the friction between them

Maglev A train that levitates above its track by use of electromagnets

Magnet An object that has a magnetic field, usually a piece of steel or similar alloy, which can attract magnetic materials and repel other magnets

Magnetic field The region around a magnet where a magnetic force can be felt

Magnetic material A substance, usually iron, steel, cobalt or nickel, that can be magnetised

Magnetic store The energy stored by the attraction or repulsion of magnetic poles

Magnetism A non-contact force between two magnets, which can attract or repel, or between a magnet and a magnetic material, which only attracts

Magnification The number of times greater an image is than its object

Mammary glands Tiny organs in a woman's breasts that produce milk

Mass How much matter there is in an object, measured in kilograms

Mechanical work Shifting energy from one store to another by a force, for example pushing or pulling an object along

Melting The change of state when a solid becomes a liquid

Menstrual cycle The monthly cycle of events that takes place in the female reproductive system

Menstruation The monthly loss of blood from the womb in girls and women; also called a 'period' or a 'menstrual period'

Metre The standard unit of length

Micro-organism An organism that is too small to see with the naked eye

Microscope An instrument used to look at very small objects

Midrib A large central vein in a leaf

Mineral An element in food needed in small amounts by the body in order to stay healthy

Mineral salt A chemical compound that contains an element essential for plant growth

Miscible Liquids that will mix with each other

Mitochondria (singular mitochondrion) Tiny structures in an organism's cell where respiration takes place

Mixture Substance made up of two or more different types of substance not chemically joined together

Molecule Two or more atoms chemically combined

Multicellular organism An organism made up of many cells

National Grid The network of pylons and wires that carries electricity around the country, from power stations to users

Natural gas A fossil fuel formed from the decay of dead sea creatures that fell to the sea floor and were covered by layers of rock

Nectar The sweet liquid produced by some flowers to attract insects

Neodymium magnet An ultra-strong type of permanent magnet

Neutralisation A chemical reaction in which an acid reacts with an alkali to form a salt and water

Newton meter A device to measure a push or pull force

Newton The standard unit of force

Niche The position in an ecosystem where a particular organism lives

Non-contact force A force between two objects that can occur at a distance, even if they are not touching

Non-renewable energy resource A finite energy resource that will run out if we keep using it

Nuclear fuel Uranium or plutonium, used in nuclear power stations to generate electricity

Nuclear store The energy stored in the nucleus of an atom, which is released in a nuclear power station

Nucleus (plural nuclei) The part of an organism's cell that contains genetic information to control the cell

Obese When someone is very overweight

Objectivity If a scientist is objective, they do not let their opinions affect their findings

Oesophagus Muscular tube that moves food from the mouth into the stomach

Oestrogen Hormone produced by the ovaries in girls and women that controls the development of secondary sexual characteristics

Ohm (Ω) The unit of electrical resistance

Omnivore An animal that eats both plants and animals

Organ Part of an organism made up of different tissues working together

Organ system A body system made up of different organs working together

Organism A living thing, made up of several organ systems working together

Origin The point (0,0) on a graph

Ovary In animals, the organ in the female reproductive system that releases eggs; in plants, the part of a carpel that surrounds the ovules and forms the fruit after fertilisation

Oviduct The tube in the female reproductive system that carries an egg cell from the ovary towards the womb (uterus), and where fertilisation occurs

Ovulation The release of an egg from an ovary

Ovule The part of a flower's carpel containing the female sex cell; it forms a seed after fertilisation

Palisade layer The main photosynthetic layer in the upper part of a leaf

Pancreas The organ associated with the digestive system that produces enzymes; it also produces the hormone insulin, which controls blood sugar levels

Parallel circuit An electrical circuit with two or more different routes for the current to take; the sum of the currents in all of these loops equals the power supply current

Paramecium A unicellular organism that lives in fresh water

Particle theory A theory that explains the properties of matter based on substances being made up of tiny particles

Payback time The time it takes for your savings to equal the cost of installing the money-saving measure, i.e. the time to make back the money you spent on it – after this time, you are in profit

Peer review Where scientists (or students) show each other their work to check that they have done things properly

Penis The male organ that puts sperm into the vagina during sexual intercourse; also carries urine out of the body

Period The bleeding that happens once a month in girls and women due to the breakdown of the womb lining; also called menstruation

Pesticide A chemical used on crops to kill pests

Photosynthesis The process by which plants make food and oxygen using light energy, water and carbon dioxide

pH scale A scale which runs from 1 to 14 and measures the acidity or alkalinity of a substance

Phytoplankton Microscopic plants that live in water

Placenta The organ that provides food and oxygen to a developing foetus from the mother's blood

Plastic deformation The permanent stretching or squashing of an object by a force acting on it; the amount of deformation is not proportional to the force, and it is permanent

Poles The two regions of a magnet where the magnetic force is strongest, labelled north and south, short for north-seeking pole and south-seeking pole

Pollen grain The part of a flower that contains the male sex cell

Pollen tube The tube that grows from a pollen grain down the style of a flower to the ovule carrying the male sex cell

Pollination The transfer of pollen from an anther to a stigma

Population The number of organisms of one species in an area

Potential difference (p.d.) The energy that a battery or cell gives to the charge which it pushes around a circuit

Power The rate at which energy is transferred, i.e. the number of joules transferred per second

Power station Where energy is released from an energy resource and transferred to the user as electricity

Precision A measure of how close your results are to each other when repeated. The closer your results, the better the precision.

Predator An animal that hunts and feeds on other animals (its prey)

Prediction A reasoned suggestion as to what the outcome (of an experiment) might be

Pregnancy When a woman has a baby growing inside her

Prey An animal that is hunted and eaten by another animal (a predator)

Producer An organism (usually a green plant) at the start of a food chain that produces the energy and nutrients for the rest of the organisms in the chain

Products The new chemical substances formed in a chemical reaction

Properties Features of a chemical substance or material

Proportional One quantity changing at the same rate as another, i.e. if the first quantity doubles, so does the second

Protein A compound needed by the body for growth and cell repair

Pseudopodia Outgrowths created in *Amoeba* by pushing out part of the cell into a finger shape, used to move through water and to engulf food particles; also called 'false feet'

Puberty The time when a child's body starts to change into an adult body due to the reproductive system becoming active and producing hormones; occurs around the ages of 9 to 15 years

Pumped storage Using excess electricity to pump water up to a high reservoir; when more electricity is needed, this water can be released through turbines to generate electricity again

Pure Made up of a single substance

Pyramid of numbers A diagram showing the number of organisms at each level in a food chain

Quadrat A sampling square used to estimate the number of plants in a habitat

Radiation The transfer of heat as a wave (infrared radiation), without need for particles; may also refer to nuclear radiation

Radioactive Describes substances that spontaneously give off potentially harmful nuclear radiation from their atomic nuclei

Radioactive waste The waste products of a nuclear power station, which continue to give off radiation for hundreds of thousands of years

Random error An error that can give too low or too high a reading, and which is compensated for by averaging a set of results

Reactants The starting chemical substances in a chemical reaction

Relative atomic mass The mass of an atom in terms of the total number of protons and neutrons in its nucleus

Relative formula mass The sum of all relative atomic masses of each atom in a compound

Renewable energy resource An energy resource that will never run out

Resistance How strongly a component opposes the flow of a current through it; found by dividing the potential difference (p.d.) across the component by the current flowing through it

Resolution is the smallest unit that can be distinguished on the equipment scale. For example, the number of divisions on a ruler or the number of decimal places on a digital scale.

Respiration The chemical reaction that releases energy from food

Rickets A disease caused by a shortage of vitamin D in the diet

Salivary glands Small organs in the mouth that produce saliva, which contains an enzyme

Sampling technique A method used by ecologists to collect, identify and estimate the numbers of organisms in a habitat

Sankey diagram A flowchart used to show the total energy input of a device, its useful energy output, and its wasted energy; the thickness of each arrow represents the proportion of energy flowing along each path

Saturated solution A solution that can dissolve no more solid at that temperature

Scrotal sac The structure in the male reproductive system that holds the testes outside the body

Scurvy A disease caused by a shortage of vitamin C in the diet

Secondary sexual characteristics The physical changes that happen to the body during puberty

Seed dispersal The spreading of seeds away from the parent plant by wind, animals, water or the explosive mechanisms of a seed pod

Self-pollination The transfer of pollen (for example by an insect) from the anther of a flower to the stigma of the same flower

Series circuit An electrical circuit with only one route for the current to take; the same current flows through every component in a series circuit

Sexual intercourse When a man and a woman have sex (make love); it sometimes results in the woman becoming pregnant

Sluice gates The panels that open or close to control the flow of water through the turbines in a hydroelectric power station

Small intestine Part of the digestive system where digestion is completed and the products are absorbed into the blood

Soft magnetic material A material such as iron that can be used to make a temporary magnet

Solar energy Using light from the Sun to generate electricity in photovoltaic cells; or heating water in solar panels by using the Sun's radiation

Solar panels Roof panels that absorb energy from the Sun to heat water, rather than generating electricity

Solar photovoltaic (PV) cells Roof panels that generate electricity directly from sunlight, rather than heating water (also just called solar cells)

Solubility The maximum amount of solute that will dissolve; measured as g/cm^3

Soluble Describes a solid that dissolves

Solute A substance that dissolves

Solution A mixture of a solute dissolved in a solvent

Solvent A substance in which a solute dissolves

Specialised cell An organism's cell that is adapted for a particular function

Species One particular type of organism; organisms of the same species can reproduce to produce fertile offspring

Specimen A sample of cells taken from an organism or a sample of a material

Speed The distance an object travels per unit time, usually measured in m/s, or km/h, or mph

Sperm The male gamete (sex cell)

Sperm tube One of the two tubes that carry sperm from the testes to the urethra; also called the vas deferens

Spongy layer A layer of loosely packed cells in a leaf below the palisade layer

Stamen The male part of a flower

Stigma The part of the carpel where pollen lands

Stoma (plural stomata) A hole/pore found mainly in the lower surface of a leaf which allows gases to enter and leave the leaf

Stomach The muscular sac where food is digested

Stopclock A device to measure the amount of time that passes during an experiment

Streamlining Giving an object a slim and pointy shape to reduce the air (or water) resistance when it moves

Style The part of a flower's carpel that holds up the stigma

Subatomic Smaller in size than the atom

Sublimation The change of state when a solid changes directly into a gas

Support force The upwards force of a surface on an object that rests upon it; this force balances the downwards force of gravity on the object, and so keeps it stationary

Surface tension The force between the water molecules on the surface of water that makes it act like a skin; this enables some insects to walk on water

Surrogate mother A woman who carries a foetus in her womb for an infertile couple

Systematic error An error that always gives results which are too high (or always too low); caused by incorrect measuring technique or a poorly calibrated instrument

Tension A contact force that occurs when a string, rope or elastic band is stretched, which opposes the stretching force

Terminal speed The maximum speed of an object; occurs when its forwards force is balance by air resistance

Testes (singular testis) Organs in the male reproductive system that produce sperm cells

Testosterone A hormone produced by the testes in males that controls the development of secondary sexual characteristics in boys

Thermal decomposition A process of splitting compounds by heating

Thermal store The energy of a substance due to the random motion of its particles

Thermometer A device to measure the temperature of a substance

Tidal barrage A dam built across a river mouth which uses the tidal motion of the water through turbines to generate electricity

Tissue A group of similar body cells working together to do the same job

True value The value of a measurement if it were made with no error, uncertainly or inaccuracy

Turbine A large rotating machine that resembles a jet engine. When steam from a power station's boiler passes through it, it spins a generator to make electricity

Turbulent flow An irregular flow of water or air around an object travelling through it, which creates a lot of drag

Ultrasound scan A scan using ultrasound waves to create an image of a baby or an organ inside the body

Ultraviolet radiation Electromagnetic radiation with a wavelength just outside the visible spectrum, beyond violet

Umbilical cord A cord made up of blood vessels that connects a foetus to the placenta

Unbalanced force A force that is not cancelled out by another force, giving a resultant greater than zero

Uncertainty The amount that your answer could be above or below (+ or −) the true value

Unicellular organism Organism made up of one cell only

Unit What a quantity is measured in, e.g. length is measured in the unit metres, and time is measured in the unit seconds

Universal indicator A mixture of dyes, which show a range of colours and indicate the pH of a substance

Upper epidermis The top layer of cells in a leaf

Upthrust The upwards force of a fluid on an object within it, e.g. the force that keeps a boat afloat in water or a helium balloon afloat in air

Urethra The tube that carries urine out of the body from the bladder; in a man it also carries sperm during sexual intercourse

Uterus Where a baby develops inside a woman's body; commonly called the womb

Vacuole The large structure in a plant cell that is filled with sap and keeps the cell (and therefore the plant) firm

Vacuum A region with no matter in it at all, not even air

Vagina Where sperm are deposited in the woman during sexual intercourse and where menstrual blood flows during a period

Validity Results are valid if the outcome is only affected by one variable

Variable A measurement taken in an experiment; the independent variable is changed to see what effect this has on the dependent variable; the control variables are not changed, so that they do not affect the outcome

Variation Differences between organisms

Variegated leaves Leaves that are multicoloured

Vertebrate An animal with a backbone

Villi Finger-like projections lining the small intestine that increase the surface area for the absorption of food molecules

Vitamin A compound needed in small amounts by the body in order to stay healthy

Volcanologist A scientist who study volcanoes

Voltage Another name for potential difference (p.d.)

Voltmeter A device used to measure potential difference (voltage)

Weight The force of gravity on an object, measured in newtons

Wind energy Taking energy from the kinetic store of moving air and using it to turn turbines, to generate electricity

Wind farm A collection of wind turbines in one place, generating electricity together

Wind turbine A device for using the wind's kinetic store of energy to generate electricity

Womb Where a baby develops inside a woman's body; also called the uterus

Work done See Mechanical work

Zero error When an instrument does not read zero when it should (e.g. with no mass on the scales)

Zooplankton Microscopic animals that live in water

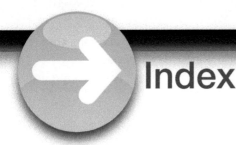

Index

Free online extras

For each topic you can find free:

- Hints for all the Questions in this book
- Extended glossary definitions and examples
- Key word tests to test your knowledge of key vocabulary.

Scan the QR codes below for each topic.

Alternatively, you can browse to www.hodderplus.co.uk/scienceprogress.

→ How to use the QR codes

To use the QR codes you will need a QR code reader for your smartphone/tablet. There are many free readers available, depending on the smartphone/tablet you are using. We have supplied some suggestions below, but this is not an exhaustive list and you should only download software compatible with your device and operating system. We do not endorse any of the third-party products listed below and downloading them is at your own risk.

- for iPhone/iPad, search the App Store for Qrafter
- for Android, search Play Store for QR Droid
- for Blackberry, search Blackberry World for QR Scanner Pro
- for Windows/Symbian, search the Store for Upcode

Once you have downloaded a QR code reader, simply open the reader app and use it to take a photo of the code. You will then see a menu of the free resources available for that topic.

→ Biology

Topic 1 Cells and tissues

Topic 2 Reproduction

Topic 3 Environment and adaptation

Topic 4 Variation and classification

Topic 5 Photosynthesis

Topic 6 Food and digestion

Chemistry

Topic 1 Particles	Topic 2 Atoms and elements

Topic 3 Acids and alkalis	Topic 4 Pure and impure substances

Topic 5 Simple chemical reactions	Topic 6 Compounds

Physics

Topic 1 Energy transfers	Topic 2 Forces and effects

Topic 3 Electricity	Topic 4 Energy resources

Topic 5 Magnets and electromagnets	Topic 6 Motion

Acknowledgements

The Publisher would like to thank the following for permission to reproduce copyright material:

p.6 © Biodisc/Visuals Unlimited/Getty Images; **p.8** *t* © Nozomu Takeuchi, Chiba University, *b* © PHOTOTAKE Inc./Alamy; **p.9** © Clouds Hill Imaging Ltd./Corbis; **p.10** *t* © Dr. Stanley Flegler/Visuals Unlimited/Science Photo Library, *bl* © Pasieka/Science Photo Library, *br* © Power and Syred/Science Photo Library; **p.11** *t* © John Durham/Science Photo Library, *b* © Dr Keith Wheeler/Science Photo Library; **p.12** *t* © Roland Birke/Photolibrary/Getty Images, *m* © Wim van Egmond/Visuals Unlimited/Science Photo Library, *b* © Russell Kightley/Science Photo Library; **p.15** © Ed Reschke/Photolibrary/Getty Images; **p.16** © ArVis – Fotolia; **p.17** © Mauro Fermariello/Science Photo Library; **p.18** © goodluz – Fotolia; **p.20** © Rosmarie Wirz/Flickr/Getty Images; **p.21** *t* © Francis Leroy, Biocosmos/Science Photo Library, *b* © PHOTOTAKE Inc./Alamy; **p.22** *t* © Ge Medical Systems/Science Photo Library, *b* © PHOTOTAKE Inc./Alamy; **p.23** *t* © Dr G. Moscoso/Science Photo Library, *b* © James Stevenson/Science Photo Library; **p.24** Gabriel Blaj – Fotolia; **p.26** © Sergey Lavrentev – Fotolia; **p.27** © Clouds Hill Imaging Ltd/Science Photo Library; **p.28** *l* © Elena Elisseeva/Dreamstime.com, *r* © M.studio – Fotolia; **p.29** *l* © Tim UR – Fotolia, *m* © andersphoto – Fotolia, *r* © Anna Kucherova – Fotolia; **p.30** *t* © M & J Bloomfield/Alamy, *bl* © Paul Broadbent/Alamy, *br* © Brian Jackson – Fotolia; **p.31** *t* © Scott Camazine/Science Photo Library, *m* © John Kaprielian/Science Photo Library, *b* © Sue Kennedy (rspb-images.com); **p.32** © Gustoimages/Science Photo Library; **p.34** *from t to b* © James Thompson/Alamy, © Kautz15 – Fotolia, © Rachel Husband/Alamy, © Suzanne Long/ Alamy; **p.35** *t* © Sylvie Bouchard – Fotolia, *b* © wusuowei – Fotolia; **p.36** © Vidady – Fotolia; **p.37** © Mel Watson/GAP Photos/Getty Images; **p.38** *t* © Nature Picture Library/Alamy, *b* © Nigel Cattlin/FLPA; **p.39** *t* © Nadezhda Bolotina – Fotolia, *b* © blickwinkel/Alamy; **p.40** *t* © bheka – Fotolia, *m* © david hughes – Fotolia, *b* © derschnelle – Fotolia; **p.44** *t* © Peter Scoones/Science Photo Library, *b from l to r* © Erni – Fotolia, © Electric Fly Co – Fotolia, © Medical-on-Line/Alamy, © blickwinkel/Alamy, © nico99 – Fotolia; **p.46** *t* © Frans Lanting, Mint Images/Science Photo Library, *bl* © Jasmin Merdan – Fotolia, *br* © Piotr Marcinski – Fotolia; **p.47** © Tay Rees/Iconica/Getty Images; **p.48** © goodluz – Fotolia, *b* © Margot Granitsas/Science Photo Library; **p.49** © Bernie Pearson/Alamy; **p.50** *t* © michaeljung – Fotolia, *b* © Nigel Cattlin/FLPA; **p.51** *t* © Farina3000 – Fotolia, *b* © Louise Gubb/Corbis Saba; **p.52** © Janine Wiedel Photolibrary/Alamy; **p.53** © Michael Howes/GAP Photos/Getty Images; **p.54** © Cultura Creative (RF)/Alamy; **p.55** *t* © RGB Ventures LLC dba SuperStock/Alamy, *b* © Ralph White/Corbis; **p.56** © Tuul/hemis.fr/Getty Images; **p.58** *t* © Grafvision – Fotolia, *b* © Dr. Keith Wheeler/Science Photo Library; **p.59** © Power and Syred/Science Photo Library; **p.60** *t* © Kenishirotie – Fotolia, *bl* © Paul Murphy – Fotolia, *br* © Nigel Cattlin/FLPA; **p.63** © Maximilian Stock Ltd/Science Photo Library; **p.64** *t* © Philippe Plailly/Science Photo Library, *m* © robynmac – Fotolia, *b* © Elena Schweitzer – Fotolia; **p.65** *t* © Alen-D – Fotolia, *b* © ia_64 – Fotolia; **p.66** *t* © Adam Hart-Davis/Science Photo Library, *bl* © Science Photo Library, *bm* © Martyn F. Chillmaid/Science Photo Library, *br* © Andrew Lambert Photography/Science Photo Library; **p.67** © Andrew Lambert Photography/Science Photo Library; **p.68** © Deep Light Productions/Science Photo Library; **p.69** © Eye of Science/Science Photo Library; **p.72** © michaeljung – Fotolia; **p.74** *t* © AntonioFoto/Shutterstock, *bl* © Sebastian Duda – Fotolia, *bm* © Dinadesign – Fotolia, *br* © Can Balcioglu – Fotolia; **p.76** © PA Archive/Press Association Images; **p.78** *t* © Mansell/Time Life Pictures/Getty Images, *m* © auremar – Fotolia, *b* © bitter...₀ - Fotolia; **p.79** *t* © DEX Images Images/ Photolibrary Group Ltd/Getty Images, *b* © kavring – Fotolia; **p.80** © Joel Arem/Science Photo Library; **p.82** © Peter Gudella/Shutterstock; **p.84** © matamu – Fotolia; **p.86** © Science & Society Picture Library/Getty Images; **p.88** © David Lyons/Alamy; **p.92** © Christopher Dodge – Fotolia; **p.94** © Science & Society Picture Library/Getty Images; **p.95** © Charles D. Winters/Science Photo Library; **p.96** © Eliza Snow/E+/Getty Images; **p.97** © Robert Brook/Science Photo Library; **p.98** *t* © skynet – Fotolia, *b* © Deymos – Fotolia; **p.102** © Damien Gus – Fotolia; **p.104** *t* © crisserbug/E+/Getty Images, *b all* © sciencephotos/Alamy; **p.106** © olavs silis/Alamy; **p.108** © Robert Estall/Corbis; **p.109** © Artem Merzlenko – Fotolia; **p.110** © Kruwt – Fotolia; **p.113** © sciencephotos/Alamy; **p.114** © E.R.Degginger/Science Photo Library; **p.116** © Martyn F. Chillmaid/Science Photo Library; **p.118** © Charles D. Winters/Science Photo Library; **p.120** © Science Photo Library/Getty Images; **p.124** © Charles D. Winters/Science Photo Library; **p.125** © Nneirda – Fotolia; **p.126** © Science Photo Library/Alamy; **p.128** © Charles D. Winters/Science Photo Library; **p.130** © Tyler Olson – Fotolia; **p.134** *t* © Purestock/Alamy, *b* © Martyn Goddard/Corbis; **p.135** © Alan Stockdale – Fotolia; **p.136** © RTimages – Fotolia; **p.138** © Red Bull Stratos/Red Bull Content Pool; **p.139** © picturesbyrob/Alamy; **p.140** *t* © Jeffrey Blackler/Alamy, *m* © Rosemary Roberts/Alamy, *b* © joefoxfoodanddrink/Alamy; **p.142** *tl* © Vivid Pixels – Fotolia, *tr* © JackF – Fotolia, *bl* © bertys30 – Fotolia, *br* © starekase – Fotolia; **p.143** © Martin Green – Fotolia; **p.144** © INSADCO Photography/Alamy; **p.146** *from t to b* © Carolyn Franks – Fotolia, © Mr Twister – Fotolia, © dvande – Fotolia, © Juice Images/Alamy; **p.148** *tl* © jonnysek – Fotolia, *tr* © Lsantilli – Fotolia; *b* © joegast – Fotolia; **p.150** © DGB / Alamy; **p.152** © kubais – Fotolia; **p.153** © Photobac/Shutterstock; **p.156** © Doug Martin/Science Photo Library; **p.158** © kalafoto – Fotolia; **p.160** © aeropix/Alamy; **p.161** © design56 – Fotolia; **p.162** *t* © Len Green – Fotolia, *b* © Xiaomei Chen/The Washington Post via Getty Images; **p.163** © Alistair Cotton – Fotolia; **p.164** © Anton Prado PHOTO – Fotolia; **p.165** © Deyan Georgiev Photography/Alamy; **p.166** © yanlev – Fotolia; **p.168** © Robert Neumann – Fotolia; **p.170** © Les Gibbon/Alamy; **p.172** © Paylessimages – Fotolia; **p.174** © Tony Craddock/Science Photo Library; **p.175** © Patrick Eden/Alamy; **p.176** © natalie willetts – Fotolia; **p.178** *t* © Sly – Fotolia, *b* © Werner H. Müller/CORBIS; **p. 180** © Alex Bartel/Science Photo Library; **p.182** *t* © TTL/Photoshot, *b* © Andrew Lambert Photography/Science Photo Library; **p.186** © Paul Hebditch/Alamy; **p.188** © DBURKE/Alamy; **p.190** *t* © PA Archive/Press Association Images, *b* © NASA - Yuri Arcurs - Fotolia; **p.191** © Erich Schrempp/Science Photo Library; **p.192** *t* © Tami Chappell/Reuters/Corbis, *ml* © dreamnikon – Fotolia, *mr* © SSilver – Fotolia, *b* © StockShot/Alamy; **p.194** © Matthias Oesterle/Alamy; **p.195** © Matthias Oesterle/Alamy.

t = top, *b* = bottom, *l* = left, *r* = right, *m* = middle

Every effort has been made to trace all copyright holders, but if any have been inadvertently overlooked, the Publisher will be pleased to make the necessary arrangements at the first opportunity.

214